Murray River Country

An ecological dialogue with traditional owners

Jessica K Weir

Aboriginal Studies Press

First published in 2009
by Aboriginal Studies Press

© Jessica K Weir 2009

All rights reserved. No part of this book may be reproduced or transmitted in any form or by any means, electronic or mechanical, including photocopying, recording or by any information storage and retrieval system, without prior permission in writing from the publisher. The Australian *Copyright Act 1968* (the Act) allows a maximum of one chapter or 10 per cent of this book, whichever is the greater, to be photocopied by any educational institution for its education purposes provided that the educational institution (or body that administers it) has given a remuneration notice to Copyright Agency Limited (CAL) under the Act.

Aboriginal Studies Press
is the publishing arm of the
Australian Institute of Aboriginal
and Torres Strait Islander Studies.
GPO Box 553, Canberra, ACT 2601
Phone: (61 2) 6246 1183
Fax: (61 2) 6261 4288
Email: asp@aiatsis.gov.au
Web: www.aiatsis.gov.au/aboriginal_studies_press

National Library of Australia
Cataloguing-In-Publication data:

> Author: Weir, Jessica Kate, 1971–
>
> Title: Murray River country : an ecological dialogue with traditional owners / Jessica K Weir.
>
> ISBN: 9780855756789 (pbk.)
>
> Notes: Includes index. Bibliography.
>
> Subjects: Aboriginal Australians — Murray River Region (N.S.W.–S. Aust.) Murray River Region (N.S.W.–S. Aust.) Murray River (N.S.W.–S. Aust.) — Environmental conditions.
>
> Dewey Number: 333.91620994

Printed in Australia by Ligare Pty Ltd

Front cover image: Nici Cumpston, Flooded Gum, Katarapko Creek, Murray River National Park, 2007. Giclée print on canvas, hand coloured with watercolours and pencils, edition 5/10, 75 x 220 cm. Image © artist, courtesy Gallery Smith, Melbourne. Collection of Flinders University Art Museum, Adelaide.
Back cover image: The beauty of a dead river red gum forest, Lake Mulwala on the Murray, New South Wales, upstream from the Barmah-Millewa forest. Image © Jessica K Weir.

Foreword

The important task Jessica Weir set for herself in writing *Murray River Country: An ecological dialogue with tradition owners* is this: How do you explain in a world prolific in human interventionism, controllers and the 'moderns' that the voice of traditional owners is relevant in today's world? What the First Nations, the Traditional Owners and their Elders have been saying over the last 70 years is that the white man's actions in the damming and usage of our rivers and waters is having a devastating effect and is in fact destroying our rivers and wetlands. We want to stop this continuing in the future. We want to be part of the solution. But in order for our voice to be heard there needs to be a real commitment to understanding and a deconstruction of those failed notions which create separateness between people and nature.

The Dreaming stories, the places of magic and creation, give meaning and essence to the First Nations. The ritual of everyday life for my river peoples is as rhythms of a song passed on from one generation to the next, each being interpreted and interwoven within the cycle of life of which water is life's blood. We belong to a particular part of the world and all living and non-living things within it are related. The hunting and gathering of our traditional foods and medicines is not a random event; we believe all life is reciprocal and is influenced by our connections and actions, each season evokes knowledge and is adapted to and responsive to any changes. Even though the impact of 'water management systems' on the lives of First Nations of the Murray River has been devastating, we do not believe it is to our demise. Indeed, our identity as Traditional Owners will prevail for we will adapt and survive as we have over many thousands of generations.

Murray River Country speaks to the potential for creating a new dialogue between the First Nations and new Australians in coming to terms and in understanding each other. This dialogue is reliant on being honest to the devastation past theories have had on our water and the river system and in changing false concepts and practices, as this is essential for not only the continuance of the Murray River system but of life itself. I am sure that I can speak for members of the Murray Lower Darling Rivers Indigenous Nations when I say how we are proud to have been part of this book and to the voice Jessica so eloquently gives to the Traditional Owners and their Elders from its inception to realisation.

Monica Morgan
Yorta Yorta woman

MLDRIN [the Murray Lower Darling Rivers Indigenous Nations alliance] is really putting the Indigenous issues upfront where everyone can see where they are. They are not swept under the table through some bureaucrat [saying], 'This happened years ago, let's forget about it.' MLDRIN is bringing it all out into the open where it should be. Let everyone see what is really going on with the Indigenous Nations.

Ramsay Freeman, Wiradjuri Elder

Itself irrigated with floods of communicative relationships, multi-partner connectivity, and natural–cultural flows that a thirsty Murray River country cries out for, this book richly waters my soul. Place, country, and care are at the heart of this wise book, which is so astutely responsive to the diverse, active Aboriginal individuals and nations of the Murray–Darling Basin, as well as to ecological sciences and ontoepistemologies of complex wholes and the coalitions of people both relearning and inventing practices of care and knowledge that bind mind, heart, and action so needed in salted times. Like the Central Valley of California near where I live, where vast rivers and wetlands have been engineered to produce a precarious and poisoned breadbasket for settler empires, the Murray–Darling Basin cries out for new practices of care from all of its people. Weir's book gives me hope that these blasted places and the lives of so many species, human and not, might again be whole, in new ways and old.
Donna Haraway, History of Consciousness Department,
University of California at Santa Cruz

Weir's originality is innovative and inspirational. She captures the MRC Indigenous people's holistic approach in reading the ecological statements of managing water and the benefits of this for everyone and the MRC's ecology.
Payi-Linda Ford, School of Australian Indigenous Knowledge Systems,
Charles Darwin University

Weir demonstrates that there is only one narrative and it encompasses both the claims of the water managers and their critics; both the settler and Indigenous narratives.
Richie Howitt, Department of Environment and Geography,
Macquarie University

This is a really positive book with original and creative suggestions about managing water using Aboriginal and non-Aboriginal understandings together.
Libby Robin, Fenner School of Environment and Society,
Australian National University

Contents

Foreword	iii
Illustrations	viii
Preface	xi
Acknowledgments	xiii
Shortened forms	xv
Interviewees	xvi
1. Narratives and their relations	1
2. Water management in the Murray–Darling Basin	26
3. Connectivity, loss and resilience	47
4. Setting the negotiation table	66
5. Murray Lower Darling Rivers Indigenous Nations	91
6. 'Deplete, destroy, depart?'	118
References	149
Index	164

Illustrations

Unless otherwise stated all illustrations are courtesy of the author

River Murray System	8
Sign next to an inlet regulator on Frenchman Creek	11
The Murray–Darling Basin	28
The Moira Channel	32
Monthly flow volumes on the Murray River	35
Yellowin Bay emerges at Blowering Dam	63
Murray Lower Darling Rivers Indigenous Nations	93
MLDRIN delegates and observers, July 2004, Deniliquin, New South Wales	98
Community Elders, October 2004, Dubbo, New South Wales	99
MLDRIN and the Murray–Darling Basin Commission, 24 March 2006, Wonga Wetlands, New South Wales	102
Wiradjuri Elders	103
The six Significant Ecological Assets	107
'Barmah Forest' by Lin Onus	136
Dead river red gums in outback New South Wales	138
The Victoria River	139
The 'former' floodplain next to the Victoria 'channel'	139
The Pink Lake, South Australia	142
Signage at Pink Lake, South Australia	142

Plates between pages 80–81
Yellowin Bay emerges at Blowering Dam, New South Wales
Flat plains in New South Wales
A dead river red gum forest, Lake Mulwala, New South Wales
Young river red gums in the Werai forest, near Deniliquin,
    New South Wales
The Murray River, Redcliff, Victoria
Fivebough Wetland, near Leeton, New South Wales
The Murray River, Wentworth, New South Wales
Carpark Reconciliation, mural, 2000, Meningie, South Australia
A midden next to the Murray River, Redcliff, Victoria

Plates between pages 112–113
The one-time water pumping station, Lake Alexandrina,
    South Australia
The Pink Lake, South Australia
The Murrumbidgee River near Yass, New South Wales
Psyche Bend Lagoon, near Mildura, Victoria
Lee Joachim's children in the Barmah Forest, Victoria
Dead river red gums line a 'regulated' creek in
    outback New South Wales
The Victoria River, now a 'channel' lined with dead river red gums
The 'former' floodplain next to the Victoria 'channel'
'Barmah Forest' by Lin Onus

Preface

The common desire to respond to ecological devastation, shared by both the Indigenous people working to support river health and the bureaucrats charged with environmental management, is stymied by the implicit assumptions the advocates for each position bring to the engagement. This is evident in water management and policy-making that assumes ecology and economy are positioned in oppositional relationships. We must acknowledge the artificial boundaries in these debates in order to open up space for dialogue, create new grounds for policy-making, and expand our capacity to address water mismanagement.

I wrote this book against a background of enormous ecological devastation in the Murray–Darling Basin, devastation that has brought environmental issues to the forefront of political agendas. I take as my departure point the unjust historical erasure of Aboriginal peoples' sovereign territories, reported so well elsewhere (for example, Attwood & Foster 2003; Brennan et al. 2005; Dodson & Strelein 2001; Clark 1995; Elder 1988; Gammage 1986; Goodall 1996; Pascoe 2007; Read 1988; and Strelein 1998). From here, I focus on how Aboriginal people negotiate water management institutions in contemporary Australia.

These negotiations are brought to life in the Murray–Darling Basin by the voices of traditional owners who have formed an alliance: the Murray Lower Darling Rivers Indigenous Nations (MLDRIN). In 1998 this alliance of traditional owners began mobilising to promote their voice in water management. The alliance covers the southern part of the Murray–Darling Basin where the governments of Victoria, New South Wales and South Australia have jurisdiction. Here, a network of rivers — the Murrumbidgee, Lower Darling, Macquarie, Edward, Goulburn, Broken and others — all eventually flow into the Murray and then out to the Southern Ocean. Access to the rivers has always been crucial for traditional owners to maintain connections with 'country' — the term they use for their land, whether it is formally recognised in Australian property law or not. In twenty-first century Australia the traditional owners live lives interconnected with non-Aboriginal people: in part distinct, in part shared, but always changing.

The research that forms the basis of this book was only possible because a research agreement was successfully negotiated and maintained between myself and MLDRIN. Since 2003 to the time of publication I attended MLDRIN meetings and events, visited important places in country and met with people, often in their homes. In 2004 I conducted semi-structured interviews with 13 traditional owners involved with MLDRIN (a list of those interviewed can be found on page xv). Their voices are threaded throughout this book, and they share with you both their profound sorrow for what they are losing, and their strategies for change. Listen carefully to the traditional owners' voices; they have inherited a knowledge tradition that emphasises connections with ecological life.

Please note that in the work that follows I have used a number of words and terms in a very specific sense. Among these words are 'traditional owners', 'governments' and 'modern' (and the related terms 'postmodern' and 'amodern').

A 'traditional owner' is a term often used to describe Aboriginal people who have responsibilities for a certain area of land. Generally, these responsibilities are passed down to them by their ancestors and their ancestral beings. The term became popularised by the *Aboriginal Land Rights Act 1976* (Northern Territory) but is now used commonly throughout Australia. On occasion I write interchangeably about traditional owners and MLDRIN delegates but while one must be a traditional owner to be a MLDRIN delegate, the reverse is not true. There are also traditional owners who are involved in MLDRIN (the organisation) but not as MLDRIN delegates.

When I speak about 'government', I use the term to encompass political institutions, such as the various state and federal parliaments, and government bureaucracies, including departments and interstate agencies. Government actors at regional and local scales, including catchment management authorities, are part of this collective term, but I have not emphasised their role.

Finally, I use the term 'modern' to describe a type of thinking that separates the world into binaries that are placed in oppositional relationships, as understood by critics of modernity and modernism (including Latour 2001, Mitchell 2000, Braun 2002, Haraway 1992, and Scott 1998). Similarly, I use the term 'amodern' (Latour 2001, p. 47) to describe a type of thinking that does not make these 'hyper-separated' distinctions (Plumwood 2002, p. 49; Latour 2001, p. 61).

JKW

Acknowledgments

The traditional owners shared with me their vision of a healthy inland river country, and taught me to understand what was in front of me that I could not see. This book would not have been possible without that learning experience. I thank them for opening their homes to me, sharing their experiences, and working together with me on this research. I would particularly like to thank Monica Morgan for her early encouragement, and Steven Ross for his commitment to this research. I would like to thank all the Murray Lower Darling Rivers Indigenous Nations (MLDRIN) delegates at the Lake Boga meeting in 2003 for responding so positively to my request for a research relationship. The mutual trust that has developed as a result of this research agreement has been central to the research outcomes.

For their interviews I would like to thank Robert Charles, Lee Joachim, Jeanette Crew, Ramsay Freeman, Monica Morgan, Junette Mitchell, Mary Pappin, Matt Rigney, Steven Ross, Henry Atkinson and Tony Peachey. Thank you for sharing with me your experiences with the rivers and your passion for river life. I would also like to acknowledge the interviews valuably contributed by Agnes Rigney and Richard Hunter, who have both since passed away. I would like to thank many other MLDRIN delegates for their support and for welcoming me at the MLDRIN meetings, especially Tracy Hamilton, Ernie Innes, Doug Nichols, Kenny Stewart, Peter Rigney, Mary Pappin Jnr, Gary Pappin and Wayne Webster. I also thank and acknowledge the delegates Ralph Harradine and Peter Murtikos who have also passed on.

I gained valuable insights and help from the people who have engaged with MLDRIN in their different professional capacities: Warwick McDonald (formally with the Murray–Darling Basin Commission (MDBC), now with the Commonwealth Scientific and Industrial Research Organisation (CSIRO), Wendy McIntyre (formally with the MDBC, now with CSIRO), Liz McNiven (formally

ACKNOWLEDGMENTS

with the MDBC, now with the National Film and Sound Archive) and especially Neil Ward (Murray–Darling Basin Authority). For their hospitality I would like to thank Helma and Matt Rigney, Jocelyn and Ian Chegwin, Steven Ross, and Sophie Germain and Tom Holt.

My academic colleagues have been central to this lengthy journey. In particular, thank you to Debbie Rose, Deirdre McKay, Lisa Strelein and Patrick Sullivan for your generous guidance and wisdom with what started out as a PhD thesis. Special acknowledgement goes to Debbie for her leadership as my supervisory chair, and to Lisa for introducing me to Monica. I also thank Tim Bonyhady, Geoff Gray, Tim Rowse and Benjamin Smith for their encouragement when I started out on this work. Much of this book was written at the Fenner School for Environment and Society, and I thank my Fenner colleagues for both challenging and supporting me, especially Daniel Connell, Thom van Dooren, Steve Dovers, and Libby Robin. I thank the Australian Institute of Aboriginal and Torres Strait Islander Studies for the time to complete this book, and Land and Water Australia who funded my PhD scholarship. I also acknowledge the Ecological Humanities, a discussion group I am a member of which provides much stimulus, inspiration and collegiality in the complex reworking of western philosophy in response to ecological devastation.

For insightful and valued feedback on the draft, my thanks go to Deirdre McKay, Debbie Rose, Steven Ross, Lisa Strelein, Patrick Sullivan, Luke Taylor, and two anonymous reviewers. For his editorial expertise in preparing the book for publication, my appreciation goes to Mark MacLean. For other editorial assistance, I thank Lara Wiseman and Ingrid Hammer. Thank you to Karl Nissan for his expertise in producing the map of the MLDRIN nations. Thank you to the staff of Aboriginal Studies Press — Rhonda Black, Gabby Lhuede, Kim Johnston, Rachel Ippoliti and Amity Raymont — for all their efforts to make this book a reality and to do so with such enthusiasm. And thanks to Luke for your care and support throughout it all.

JKW

SHORTENED FORMS

AIATSIS	Australian Institute of Aboriginal and Torres Strait Islander Studies
ATSIC	Aboriginal and Torres Strait Islander Commission
CAC	Community Advisory Council
CCCG	Carp Control Coordination Group
CMA	Catchment Management Authorities NSW
COAG	Council of Australian Governments
CSIRO	Commonwealth Scientific, Industrial and Research Organisation
DAA	Department of Aboriginal Affairs
DEH	Australian Government Department of Environment and Heritage
DEWHA	Department of Environment, Water, Heritage and the Arts
DIPNR	Department of Infrastructure Planning and Natural Resources
DWLBC	Department of Water, Land and Biodiversity Conservation
FCG	Farley Consulting Group
MDBA	Murray–Darling Basin Authority
MDBC	Murray–Darling Basin Commission
MDBMC	Murray–Darling Basin Ministerial Council.
MLDRIN	Murray Lower Darling Rivers Indigenous Nations
MoU	Memorandum of Understanding
NCCMA	North Central Catchment Management Authority
NPA	National Parks Association of NSW
NWI	National Water Initiative
UNCESCR	United Nations Committee on Economic Social and Cultural Rights
VEAC	Victorian Environmental Assessment Council

INTERVIEWEES

Henry Atkinson, Yorta Yorta, MLDRIN delegate [7 August 2004]
Robert Charles, Wamba Wamba, MLDRIN delegate* [24 June 2004]
Jeanette Crew, Mutti Mutti, Staff, NSW Department of Environment and Climate Change [26 June 2004]
Ramsay Freeman, Wiradjuri, MLDRIN delegate [27 June 2004]
Richard Hunter, Ngarrindjeri, MLDRIN delegate* [23 July 2004]
Lee Joachim, Yorta Yorta, MLDRIN delegate* [25 June 2004]
Junette Mitchell, Barkandji, Occasional observer at MLDRIN meetings [20 July 2004]
Monica Morgan, Yorta Yorta, Former Staff, Yorta Yorta Nation Aboriginal Corporation, and Murray–Darling Basin Commission [1 July 2004]
Mary Pappin, Mutti Mutti, MLDRIN delegate [22 July 2004]
Tony Peachey, Wiradjuri, MLDRIN delegate* [29 October 2004]
Agnes Rigney, Ngarrindjeri, MLDRIN delegate* [21 July 2004]
Matt Rigney, Ngarrindjeri, MLDRIN delegate [24 July 2004]
Steven Ross, Wamba Wamba, MLDRIN Executive Officer [28 July 2004]

*no longer a MLDRIN delegate at the time of publication

CHAPTER 1

Narratives and their relations

The relations between the old nature and the new nation are tense, at times even violent.
Libby Robin, environmental historian[1]

BIG HILL

In the heat of January 2008 I drove through the inland river country, catching up with the traditional owners who had helped me to write this book. From Canberra, I drove west and south alongside the Murrumbidgee and Murray rivers to eventually arrive at the Coorong — the long narrow body of water made famous in Australia by the movie *Storm Boy*. The Coorong is protected from the wild Southern Ocean by a high row of sand dunes hundreds of kilometres long. The Coorong is a mixture of salty seawater and the fresh river water of the Murray River. The Murray flows into lakes Alexandrina and Albert, and the Coorong, as part of its journey out to sea. Scientists tell us that the Coorong is dying; the water is now saltier than the sea itself and not much can live in it. The pelicans made famous in *Storm Boy* are still present but they are no longer breeding.

Here I met up with Matt Rigney and his brother Peter, two Ngarrindjeri men who grew up next to the lakes and the Coorong. Together we drove to the top of Big Hill, the hill behind Raukkan, the former Point Macleay Mission. From the top you can look north and west over Lake Alexandrina to Mount Barker and Point Sturt. You can look south down to the small peninsula that marks the beginning of the Coorong. Matt told me about growing up at the mission and the times he spent going camping and living off the land with his father. Matt pointed out the spot where his father shot and wounded a duck, and how Matt and Peter were sent out after it. They chased that wounded duck across the shallow bed of Lake Alexandrina, way out to the deeper waters of the Murray River, and then they swam for another 40 metres before they caught it and broke its neck. He spoke about the old ladies who used to go out fishing on the peninsula; it was their favourite spot.

---

1. (Robin 2007, p.5)

Just down from where we stood was a cave where the old men used to sit. As a child, Matt would listen to the Elders talking in the Ngarrindjeri language. They told him how they relied on the pelicans to find the body when somebody drowned: when a pelican sees something in the water it circles around and swoops down, but it will rear back up if it is a human body and not a fish.

From on top of Big Hill Matt used to grab sheets of corrugated iron and slide down, splashing into the water. One time he and his friends found a *pondee*, a Murray cod, down there, dying in the shallow water with its gills full of sand. It was so big that they needed two sticks to carry it, one stick on each side threaded through the gills, and with three of them on each stick they carried it back to the camp. Much more recently, five or six years ago, Matt sat here and shot carp. This introduced fish has become present in plague numbers. On this day we all looked at how the water level had dropped, no longer reaching the foot of Big Hill.

The devastation of the Murray River is keenly felt by the many Aboriginal people I met who call the inland rivers their country. A week earlier I visited Mutti Mutti Elder Mary Pappin in the Murray River town of Mildura. She talked to me about the industry in mining the river sand. Mary asked rhetorically, 'Is this what you do when you've killed everything on top: you take out the minerals from underneath and sell them?' Mary can always be counted on for going to the heart of the matter.

A week after the Coorong I was in Echuca visiting Yorta Yorta woman Monica Morgan. Monica told stories about growing up in Echuca, skipping school and hiding under the old riverboat pier made out of large river red gum planks. When the school principal came to get them they would dive into the water and swim across the Murray, out of reach. But Monica also spoke about the present and the need for the different nation groups to come together and strengthen their cultural ties. Monica talked about how the Kulin language group have the Eagle Hawk and Crow Dreaming, whereas the Yorta Yorta have the Rainbow Snake Dreaming which is more similar to the Ngarrindjeri Dreaming. Monica wished the nations (see pp. 103–4) had more time to share and celebrate this diversity with each other, rather than always being caught up with responding to government agendas in water law, policy and management.

I also visited Yorta Yorta man Lee Joachim in Barmah, and he shared with me the land use and occupancy mapping work the Yorta Yorta are doing. As part of this mapping process the Elders have identified all the different animal and plant species they used to hunt and gather. In discussion with scientists it became apparent that the Elders had identified two fish species, previously unrecorded in scientific inventories, the *walka* and the *gilgarja*, that used to be found on the grass plains in times of flood but are now absent.

1. NARRATIVES AND THEIR RELATIONS

There are many stories that the traditional owners tell about life and death next to the Murray River. Their stories are an integral part of the Murray River and the Murray–Darling Basin. They describe an ancestral domain and a fertile homeland where the Murray River is a life force that created the life of the inland river country. The Murray–Darling Basin has also been characterised in Australian national identity in stories of progress and nation building, stories in which human skill and ingenuity have created a vast agricultural heartland. In this story flowing rivers of liquid gold have supplied the economic resources for vast amounts of agricultural produce and the country towns that have thrived with this economic growth.

The history books tell us that the spiritual homelands of the Murray River peoples have given way to the contemporary priorities of modern agricultural production, in this, Australia's agricultural heartland. But this portrayal of history is actually what I call 'modern thinking'. In modern thinking, Aboriginal peoples' stories are a narrative that is spiritual or traditional, and the nation-building narrative is one of economic growth or development. From the modern-thinkers' perspective these narratives cannot co-exist: one must be sacrificed for the other.

In the text that follows I shall show how this way of thinking is false, and, critically, disables our responses to the ecological devastation we now face in the Murray–Darling Basin. The far-reaching relationships that are sustained by water enmesh water as a resource for production *and* water as an ancestral life force. To understand these close relationships we must examine how modern thinking develops such narratives as both separate and oppositional, in sharp contrast to Aboriginal peoples' knowledge traditions. However, simply establishing Aboriginal peoples' perspectives as being oppositional to modern thinking is not my intention. I shall reveal how intertwined all our knowledge traditions are, so that we can keep that which is useful from modern knowledge, whilst also acknowledging the limitations of modern thinking. To do this we must first take some time to understand exactly what 'modern thinking' is. With that understanding, we can then analyse how modern thinking determines the intellectual space available to Indigenous people in their engagements with governments.

MODERN THINKING AND WATER MANAGEMENT

Modern thinking about nature and society was part of the inheritance that the predominantly European colonialists brought to Australia in the eighteenth century, and was mobilised in the relationships they created with Australia's freshwater rivers, lakes, creeks, swamps, ponds and wetlands. The new settlers used modern thinking to translate the country they encountered into resources for development.

For most of these settlers, Australia was an empty space available for the agricultural industry necessary to develop their young nation (Robin 2007).

Such ideas about nature and society have a relatively recent history. 'Modern' knowledge and theories were argued for by people during a period in western history called 'modernity'. Modernity is often described as beginning with the Enlightenment, a period when intellectuals in eighteenth-century Europe and North America argued in favour of reason as the basis for authority (Mitchell 2000). These 'moderns', as I call them, argued that knowledge based on science, reason and rationality had more authority than knowledge based on religion, intuition and emotion. Moderns sought, and continue to seek, to establish science and reason as the foundation of knowledge, and thus free the world from superstition and religious beliefs. The moderns argue that comprehending the world through science and reason is the only route to discover truth, and that this truth must be founded on objectivity.

The philosopher René Descartes, considered to be the first influential thinker of this period, envisioned that science and philosophy would empower men to be the masters and possessors of nature. René Descartes held a mechanistic view of the world, one in which reductive analysis could explain the whole. From his perspective, nature (including the human body) could be reduced to matter, without agency (Mathews 1994, pp. 17, 31). He argued that only humans have agency, and this is in the mind, which he reasoned was separate from the material body. Further, that animals do not have minds and thus do not reason; indeed, they do not even feel pain. Thus, humans are positioned as the only creatures capable of thought, reason, and agency.

What René Descartes did was to amplify the western knowledge tradition of dualism. In Cartesian dualism two fundamental concepts exist in opposition to each other, forming binary pairs. These binary pairs are very familiar; for example, male/female, mind/body, rationality/emotion, spirit/matter, subject/object, nature/culture or ecology/economy. In a dualistic approach, nature is objectified, posited as external to and used by the human subject.

Much has been built upon this dualistic thinking. Contemporary French philosopher Bruno Latour has argued that the separation of the world into binaries allowed the moderns to mobilise the earth's resources on an immense scale. He concluded that these modern arguments made it possible for people to 'liberate productive forces' without accounting for the 'delicate web of relations between things and people' (Latour 2001, pp. 32, 39). According to the moderns, the people who came before them are the 'premoderns', people who inhabited a world constrained by their conceptual mixing of divine, human and natural elements. It

1. NARRATIVES AND THEIR RELATIONS

was the impossibility of changing the natural order without modifying the social order — and vice versa — that obliged the premoderns to exercise great prudence in their relationships with nature. In furthering the conceptualisation of the world into binaries, particularly the separation of nature and society, the moderns created a new view of the world, one in which they could change the natural order without consequences for the social order (Latour 2001, pp. 41–3).

The enormous growth in the technological capacity of humans since the Industrial Revolution has made it possible for the proponents of modern thinking to extend their world view of binaries on a grand scale. One example of the triumph of modern thinking is the increase in large dam construction from the early twentieth century onwards. Political scientist Timothy Mitchell has described how dam builders employed Cartesian dualism in the reorganisation of relationships with the Nile River when they built the Aswan Dam in Egypt. Timothy Mitchell argued that this project was conceived and implemented in a way that involved the separation of human expertise on the one side and nature, as the river, on the other (Mitchell 2002, p. 35). The oppositional binaries mobilised by the dam builders included science/nature, material reality/human ingenuity, stonework/blueprints and objects/ideas. With all the engineering gathered in one location — the dam site — an observation point was provided for people to comprehend 'the river as a force of nature tamed by man' (Mitchell 2002, p. 36). It was possible to undertake such extensive projects because the river was viewed as matter, hence there would be no consequences that could not be managed through engineering.

The moderns were aided in their harnessing of the productive forces of nature by their perception of nature as being all the same, as defined by a universal science. Moderns argue that through scientific process it is possible to show that nature adheres to the same rules across the world (Latour 2002, p. 105). This is achieved through the observation of controlled experiments in a laboratory to 'make facts'. This 'fact-making' was first demonstrated in England by seventeenth-century natural philosopher Robert Boyle, who created an experiment with an air pump to measure the weight of air (Latour 2001, pp. 17–8, 24). By replicating the results of such scientific experiments, the moderns were able to reveal that nature is universally the same, and to simultaneously reveal that science is universal knowledge (Smith 2007, pp.78–9). Anthropologist Paul Sillitoe discusses this powerful combination (Sillitoe 2007, p.13):

> Regardless of the discoverers, whatever they uncover will be the same because nature is a constant beyond those who investigate her, such that they will all reveal the same mysteries.

Paul Sillitoe points out that, from the perspective of universal science, all the scientist is doing is simply revealing nature to us in order for us to advance our understanding of the world. But this approach encourages the negation of local difference through the creation of apparently 'objective' formulas.

This idea of universal nature was critical for constructing 'nature' as a collection of 'natural resources'. In Germany, forestry science abstracted trees into volumes of lumber that could be used by centralised government authorities. A 'fiscal forestry' was created that only valued certain parts of the forest: rows of trees producing lumber, not the moss or lichen, the forest spirits and other forest creatures. The creation of this unified forest increased the possibilities for centralised forest management (Scott 1998, pp. 12–13, 18).

Canadian geographer Bruce Braun has also shown how such universal notions of nature form part of the operation and legitimacy of the forestry industry. In British Columbia, Canada, it was necessary to disavow other notions of the forests in order for the government to claim the trees as a commodity for the nation (Braun 2002, p. 9). By representing the forest as an abstraction, the government could focus on the rational management, economic development and ecological sustainability of the forest in line with the best scientific expertise available. The forest that was 'normalised' was one that provided the optimal rotation of sustained-yield forestry; that is, a forest that has an equal distribution of trees of different ages. Sustained-yield forestry works on the principle that each year the amount of fibre removed from the forest is equal to the amount of fibre added to tree growth (Braun 2002, pp. 41, 43, 66). The other uses and values of the forest are erased by this approach.

In Australia, environmental historian Libby Robin has detailed how the confluence of science and government became the hallmark of the modern nation. Historically, the Australian Government has actively managed scientific research agendas to prioritise agricultural, forestry and soil sciences. State enquiries have funded a universal science focused on the economic potential of primary industries. This funding has always been at the expense of the distinctive sciences of Australian plants, animals and places (Robin 2007, pp. 202, 215). Colonialist perceptions of Australia as isolated and strange reinforced the new settlers' reliance on imported European ideas. The island continent was described as nature's museum, with backward and primitive plants and animals (including humans) that were living antiquities (ibid., p. 33). This was 'useless', 'rotten' country. With the exception of forestry, the new settlers found it necessary to import European species for commercial agricultural production (ibid., p. 186).

1. NARRATIVES AND THEIR RELATIONS

The settlers applied universal science to inscribe their vision on their new country (ibid., p. 206). The ancient soils were mined in the first years by nutrient-hungry European agricultural crops. Within a few years the settlers had depleted the soils, but with the application of fertilisers and with irrigation it became possible to establish an intensified agricultural livelihood. The colonial authorities planned to grow the commercial products desired by European markets. Indeed, the federation of Australia in 1901 was pursued by the colonies so that they could better engage with the international economy (ibid., pp. 214–15). Libby has argued that the new settlers recast Australia as a neo-Europe: 'even the idea of subjugating nature and forcing yields was imported' (ibid., p. 186).

The management of water resources was central to this vision, as Australia is the driest inhabited continent on earth. In pursuit of the vision of a robust agricultural economy producing for export, the south-eastern colonies, soon to become state governments, decided to regulate the supply of water to agricultural producers. In the twentieth century these governments, along with private industry and water experts, collaborated to build dams on a grand scale, aided greatly by innovations in technology and engineering. Ephemeral lakes were converted into permanent water storages, river valleys were flooded, canals and channels were dug, wetlands were drained, and a series of locks were built across the Murray River and its tributaries (Powell 1989; Crabb 1997). The intent of the dam builders was to store, control and regulate water in order to allocate water to farmers, pastoralists and townspeople. This activity was only possible by thinking of water as a commodity, thus turning it into an abstract resource for the nation rather than engaging with it as the meaningful, connected and critical life force that it is.

Today, the modern approach to water management in the Murray–Darling Basin is exemplified in the following diagram (p. 8). This map shows the River Murray System, as conceived by the Murray–Darling Basin Commission (the Commission), the institution that preceded the creation of the Murray–Darling Basin Authority in 2008. The Commission describes the River Murray System as the main course of the Murray, including its effluents and anabranches; all tributaries entering the Murray upstream of Albury; the Darling River downstream of the Menindee Lake storage; the Dartmouth and the Hume dams, Yarrawonga Weir, Lake Victoria storage, Menindee Lakes storage, and the weirs and locks along the Murray and lower Murrumbidgee; and the barrages near the Murray Mouth. Its internal business division, River Murray Water, operates and manages the River Murray System using four major storages, 16 weirs, five barrages and many other smaller structures.

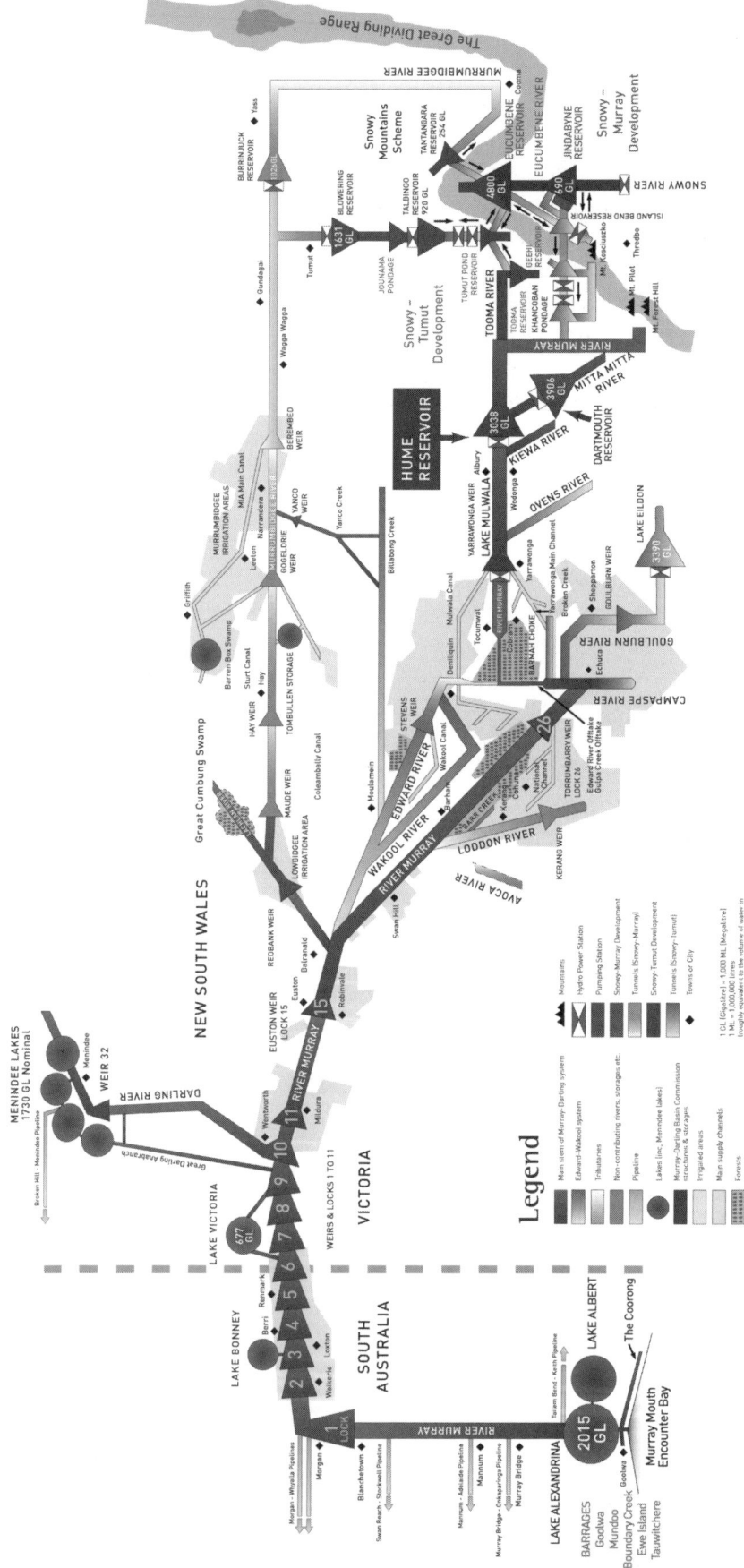

A map of the River Murray System. © Murray–Darling Basin Authority.

1. NARRATIVES AND THEIR RELATIONS

The main course of the Murray River is shown simplistically as a pipe, with a series of supporting weirs, locks and water storages that manage the movement of water along the river and out to irrigation fields. The diagram illustrates an appreciation of the Murray River as an externally constituted water resource that is moved and allocated by human ingenuity through engineering works.

The language of the Commission reveals modern thinking at work. It imagines a river that is under the control of staff working for River Murray Water. The operation of this imagined system is termed 'Running the River', as the Commission described on its web page (MDBC 2007h):

> Each day, River Murray Water staff 'run the river' by deciding on releases from storages along the River Murray and lower Darling. We release water to meet the needs of irrigators and flows for South Australia within constraints such as minimum flow requirements, dilution of salinity, maximum rates of change of water level, and capacity of the river channels. Instructions on releases are then given to staff of the State Authorities located at the various river structures …
>
> Each day, more than 350 items of data are received from points along the River Murray and tributaries to facilitate operation of the River Murray system … This data is put into a computer database as well as manually recorded on 'operations sheets'. A computer-based model uses data from the last few weeks to predict flows at key points along the River System. The model takes into account the time it takes for water to flow between points, the magnitude of diversions of tributary inflows (River Murray Water obtains forecasts from the States), irrigation water, 'loss' of water due to evaporation and seepage, and the change in volume of storages and weir pools. The model also estimates future requirements for water. After consideration of downstream demands, losses and tributary inputs, releases are determined for each major storage to ensure efficient delivery of water to downstream users. Staff at the river structures are then instructed to implement the releases.

In this text, the Commission explains how River Murray Water staff have a management solution for the different water flows of the river, by analysing data and relating it to the model they have created. These management solutions are organised into 'three modes of running the river' (MDBC 2007h):

> Supplying mode occurs for most of the irrigation season, which typically extends from the start of November each year until mid May. During supplying mode operations, the flow in the river is set to meet the demands (including the entitlement flow to South Australia) with little excess … The storing mode of river operation generally occurs during the winter and spring seasons, when the flows in the river are in excess of that required to meet diversions, water supply,

and minimum flow requirements; but which are confined within the channel ... Spilling mode occurs when flow exceeds the river's channel capacity at a point as a result of run-off generated by heavy rain. This operation can be quite complex as the flow varies as tributaries join the main stream. The channel capacity varies along the River Murray.

The Commission has further explained how these three modes relate to each other (MDBC 2007h):

> It is possible for one reach of the system to be run in one mode, while another reach is simultaneously run in a different mode. For example, a major flood in the Darling River in January may lead to spilling mode being employed for the Darling River and the River Murray downstream of Wentworth, while at the same time the River Murray upstream of the Darling Junction may be operated in supplying mode to meet irrigation and other demands.

The extensive infrastructure and technology that has been developed and built to manage and operate this colossal undertaking can be seen along rivers throughout the Murray–Darling Basin. As described by the Commission, much of this infrastructure is handled by the River Murray Water staff in their office in Canberra.

One part of this system is the operation of inlet regulators. These are operated remotely by River Murray Water staff to regulate the amount of water flowing out of a water storage unit and into an inlet. The photograph opposite shows an information sign erected at an inlet regulator at Frenchman Creek.

Frenchman Creek is in the remote south-west of New South Wales, on the back road between the Murray River towns of Mildura and Renmark. A remote part of the state, with a remotely operated inlet regulator, completes the modern dream of a remote-control river.

LIFE WITH COUNTRY

At the time of colonisation, Indigenous peoples living in what is now known as Australia had their own knowledge traditions which, until that point, had been developing with very little interaction with the few European and South-East Asian peoples who visited the continent. Today, Indigenous peoples' philosophies and knowledge are deeply engaged with modern thinking. However, Indigenous people continue to articulate their concerns in ways that require modern thinkers to examine the truth of their own assumptions.

1. NARRATIVES AND THEIR RELATIONS

A warning sign indicating gate changes, next to
an inlet regulator on Frenchman Creek,
New South Wales.

One of the most common characteristics of Indigenous peoples' knowledge, in comparison to modern thinking, is an emphasis on knowledge coming from a specific place. This place is known as 'country'. Country is profoundly important to traditional owners, who are generally the people who have inherited the rights and responsibilities to country from their ancestors and ancestral beings. For them, these are innate ties between particular people, land, law and language (Smith 2005, p.8). Theirs is a knowledge system based on relationships and connections with country, so it is often characterised as a contextual or holistic knowledge. Proponents of holistic knowledge advocate that the whole cannot be understood simply as the sum of its parts; that is, the whole is more than the sum of its parts.

In 2002 and 2003 the Commission engaged the Farley Consulting Group to meet with Aboriginal people to gauge their concerns about water management in the Murray–Darling Basin. This work was coordinated with the Murray Lower Darling Rivers Indigenous Nations (MLDRIN), a confederation of traditional owners from along the Murray River and its tributaries. MLDRIN held eight meetings with traditional owner groups, and the Farley Consulting Group held five public forums for Aboriginal people more generally.

Throughout their report, the Farley Consulting Group provided feedback in the form of direct anonymous quotes, including the following (FCG 2003, pp. 12, 17, 18, 20–1, 22):

> The healing that we use Old Man Weed for needs to be done by the River. It is the same with fish — we need to catch, cook and eat by the River. Now, we can't get clay out of the bank to coat the fish or to use on our skin — this is a big part of women's business.
>
> The land and the rivers and the people are one.
>
> It should come back to life from the reeds to the insects.
>
> It is our lifeblood.
>
> It is life.
>
> Our beliefs are entwined in the river itself.
>
> We should be able to see the cod and hear the frogs. There should be plenty of yabbies and mussels. There should be reeds, catfish and birds. The grasses should come back.
>
> The scales are now unbalanced and our Ancestors are unhappy and restless. They will stay this way until balance is restored.

The Farley Consulting Group found that Aboriginal people emphasised 'respect for country' as their main concern about their relationships with the river country. The report's authors (FCG 2003, pp. 5–6; original emphasis) summarised these concerns thus:

> The river system must be treated with respect, as it is the lifeblood of the country. If the river is in poor health, it can not provide spiritual, cultural, economic and social benefits to all those who depend on it.
>
> The basis of management of the river system must be a whole landscape approach, including all tributaries of the River Murray. The objective for management of the river's resources must be *sustainable use* with the core values of the river system preserved as a legacy for future generations.
>
> To fully respect the river and all adjoining systems, the mouth of the River Murray should be open. This can only occur if the needs of the river are respected — it effectively means increasing natural flows, bringing back native fauna and flora and eradicating introduced species such as carp and willow trees.

Three central themes about country come out of this consultation: respect, mutuality and connections. These are concerns about relationships. The relationships are held between people, rivers, clay, cod, reeds, insects, yabbies, birds, grasses, ancestors and more. These relationships are described as important for culture, healing, eating and living next to the river. Taking care of the river's needs takes care of these matters. The river has needs that should be respected. Respect is key for the continuance of these life-sustaining connections. Significantly, the Aboriginal people speak about the rivers, plants, animals and ancestors as having spirit and agency in these relationships. Rather than the mindless matter posited by the moderns, country is alive with the multiple agencies of these other beings, and the knowledge traditions of the traditional owners depend on country for perpetuity.

In these holistic understandings of relationships with other living beings, Aboriginal people keep mixing the modern binaries, combining nature with culture, human bodies with rivers. An example comes from Yorta Yorta man Lee Joachim, when he talks about the Barmah–Millewa lake on the Murray (Corowa 2006):

> To us, we consider it to be the kidney of our people. And the flooding regime that needs to come through those kidneys and out to the land, flush the land, cleanse it, so it flows on.

Lee blends kidneys and lakes together. This nature cannot be understood simply as matter. Aboriginal people describe how respecting and understanding the life and agency of the river country is important for the continuation of all life. This emphasis on intimate relationships with water is in stark contrast to modern views of country and water as physical objects for human use. Yet this does not also mean that water is not appreciated as a resource for human use. Indigenous people include diverse issues of utilitarian needs, law, economy, health and more when they talk about the importance of country. Indigenous academic Steve Kinnane has articulated country thus (Kinnane 2002, p. 25):

> All regions of country are regarded as needing human obligatory ownership to be maintained through spiritual and cultural practices. This totemic understanding of human relationships to country does not disregard land used for utilitarian purposes, but operates in a way that is restricted by overarching spiritual and cultural imperatives. The concept of country does not allow for a separation of people, land and waters. In an Indigenous vision of country, economy, spirituality, knowledge and kin are all related and interconnected. Country is not seen as being 'owned' as in the Western tradition. Rather, country is held in a reflexive, obligatory way.

The objectification and use of resources is part of Indigenous knowledge and philosophy, but this value of country does not dominate to the exclusion of all other values. Rather, resource-use decisions are made with reference to a range of other concerns, including dialogue with the ancestors (Langton 2002).

Indigenous people speak about water as a web of relations within which life, spirit and the law are connected, whereas the moderns have created a far narrower vision of water as a resource to be stored, regulated and allocated for human consumption and economic production. This narrowing of vision by the moderns has been necessary to transform water into a legible phenomenon that can be measured and calculated by governments. Ecological and social relations are so complex and varied that governments use an intellectual filter to screen out complexity so as to read, aggregate and manage ecological and social relations on a large geographic scale (Scott 1998, pp. 11, 22).

It is easy to characterise, and then dismiss, the connected perspectives of Indigenous peoples as a chaotic or undifferentiated view of the world, constricted by culture, tradition and religious beliefs, disengaged from practical concerns, and unable to be read or quantified as the moderns have been able to do. Instead, as anthropologist Debbie Rose has argued, Indigenous peoples' philosophies are not simply the view that 'everything is connected to everything', rather, everything is connected to something, and there are patterns of connections: healthy, torn, patchy and intricate (Rose 2007b). It is possible to follow connections and to make distinctions; to differentiate so as to analyse, understand, respect and objectify the world. Moreover, understanding that one's life is held in the hands of other beings, and vice versa, lays the groundwork for building ethical obligations between species, and between all sorts of life forms (Rose 1999, p. 175).

In the Northern Territory, geographer Sue Jackson has described how Indigenous people from the Daly River value their connections with water. These life-sustaining relationships are being confronted by the current push by state, territory and Commonwealth governments to allocate more river water for human use and consumption. As Malak Malak Elder Biddy Lindsay explained:

> Living water — like animal — I think of it like that — living thing. White man way — don't care for that water — how you going to live?

Wagiman Elder Paddy Huddleston also explained (in Jackson 2005, p. 141):

> That river and human being are connected. They're alive. It's killing Aboriginal life [all that water that's going to get used for farms]. The water and the trees are connected to Aboriginal life.

1. NARRATIVES AND THEIR RELATIONS

These comments show how the world is not a passive instrument or neutral surface for human activity, but is alive with the agency of many other beings, innately connected to human life. These perspectives come from knowledge traditions that are not steeped in dualism, and that have only recently been intensively interacting with modern knowledge.

POSITIONING KNOWLEDGE

In describing modern knowledge and Indigenous knowledge I have highlighted the differences in the two approaches, necessarily glossing over complexities to make a point about different philosophical traditions. But what I really want to do is show how the knowledge traditions relate to each other. To do this, we must move to understand the critiques of modernity.

Philosophers have long criticised the work of moderns such as René Descartes. The most well-known of these critics are known as the postmoderns. (Note that I use the term 'postmoderns', rather than postmodernists, to maintain their relationship to Burno Latour's 'moderns'.) Postmoderns argue that the 'grand narratives' of modern thinking smother the particular narratives of this world; that is, the narratives of different cultures and marginalised people (Arslan 1999, pp. 196, 207). Postmodern approaches show that universal, rationalist narratives, purportedly based on reason, are actually culturally specific. Thus, efforts to establish objectivity and order have to acknowledge personal and cultural biases. However, a critique of the postmoderns shows that they take this argument to its extremes: everything is discourse and nothing can be known. According to postmoderns we can only engage with a world of images. As eco-philosopher Donna Haraway has described (Haraway 1992, p. 297):

> Hyper-productionism [modernism] refuses the witty agency of all the actors but One; that is a dangerous strategy — for everybody. But transcendental naturalism [postmodernism] also refuses a world full of cacophonous agencies and settles for a mirror image sameness that only pretends to difference.

Bruno Latour has also argued that postmoderns still live within modernity even though they no longer believe in it. Postmoderns believe in the divide between the technical and material world on one hand, and the discursive world of text and language on the other (Latour 2001, pp. 46, 61). Timothy Mitchell has analysed how this relationship is embedded in the representation of things (Mitchell 2000, p. 20):

> The post-modern is typically understood as a world of images and replications that have lost touch with this supposedly original reality. The real, it is said, has been replaced by the pseudo-real. In other words, such accounts continue to assume the unproblematic nature of a distinction between the real and represented, even as they announce its historical disappearance. For this reason, most theories of post-modernity remain within the binary metaphysics of the modern.

Timothy Mitchell has argued that postmoderns still believe in the modern categories — nation, nature and economy — as representing something that really exists, even though they also believe that we can never know that reality. The categories have been repeated so effectively in modern thinking that it seems like these conceptual objects 'exist prior to any such representation' (Mitchell 2000, p. 19). We are unable to consider different possibilities whilst we are caught within the framework of unexamined representations.

Bruno Latour writes that instead of going down a path of post-postmodernism and post–post-postmodernism *ad infinitum*, we need to critically engage with the two parts of the work of the moderns: first, the work that they promote as their own; and second, their work that is concealed from them.

The first part includes the creation of binaries, such as that of humans/nonhumans, and their properties and their relations. I have described this binary creation earlier in the chapter and used the term 'modernity' narrative, although Bruno Latour describes it as 'purification'. Purification is the process of separation that allows the moderns to 'liberate' the productive forces of nature.

In the second part of Bruno Latour's theory about the moderns he describes as 'proliferation' — the exponential growth in the mixing of binaries — and this is the part that remains concealed to the moderns. He argues that the moderns think they have been successful because they have been able to effectively separate nature and society (purification), whereas their success has really come from mixing together 'much greater masses of humans and nonhumans' without ruling out any combination (Latour, 2001, p. 41). The outcome of the success of the moderns is not separation but the amplification of connections in all sorts of imaginable and unimaginable ways (proliferation). For example, as a result of modern water management in the Murray–Darling Basin, where now does the Murray River begin and end in relation to the Murray River System? Who or what is the Murray River today? Bruno Latour gives a diversity of examples, including the hole in the ozone layer, frozen embryos, and whales wearing collars fitted with radio tracking devices to illustrate his point (Latour 2001, pp. 1–2).

1. NARRATIVES AND THEIR RELATIONS

Bruno Latour also identifies a causal relationship between purification and proliferation, and notes that the moderns only credit purification as contributing to their success. The more that science is 'absolutely pure', the more it is blended with the fabric of society. This proliferation has been on an immense scale; anything is possible with the moderns' perceived power to manipulate a separate nature. But it is the inability of the moderns to identify their work of proliferation that has allowed them such scope (Latour 2001, p. 43).

Egypt's Aswan Dam, is an example of proliferation. Modern thinking informed the rationale of the project as human ingenuity overcoming nature, but the reality was a complex engagement with many actors. The dam was not a result of a calculating human agency directing towards social outcomes, but more of a series of alliances in which the humans were never wholly in control (Mitchell 2002, p. 10). Timothy Mitchell draws on arguments about representation to argue that (Mitchell 2002, p. 52):

> Ideas and technology did not precede this mixture as pure forms of thought brought to bear upon the messy world of reality. They emerged from the mixture and were manufactured in the processes themselves.

He extends our understanding of these issues by contextualising the Aswan Dam construction within the local history of irrigation. Long before the twentieth century, human, mechanical, animal and hydraulic water power had worked to channel floodwaters into a complex irrigation system of hundreds of interconnected fields. This geography was 'no more natural than it was human, and no less. Rather, it was always both.' With the construction of the Aswan Dam, 'Nature was not the cause of the changes taking place. It was the outcome' (Mitchell 2002, pp. 34–5). What the moderns succeeded in doing was mixing nature and culture on a much greater scale than before, yet at the same time they were further separating their philosophical understanding of nature and culture: these are the two modern processes of proliferation and purification. The role of other actors continued despite the modern denial of their existence, and they immediately began forming their own opinions of the engineers' vision. For example, snails and mosquitos made use of water projects to spread malaria and schistosomiasis, connecting their agency to a war, an epidemic and a famine. Recognising this agency is not to move to an understanding where events are determined by the snails, mosquitos or rivers, but to acknowledge the inseparability and multiplicity of these histories and actions (Mitchell 2002, pp. 23, 27, 34, 53).

According to Bruno Latour, it is only possible for the moderns to undertake their work of purification if they ignore the proliferation of amplified connections that their work generates. The work of the moderns is effective because it assumes the logic of a clear separation. The moderns can proceed because of what they do not, or cannot, see. Where the binaries are embedded together, the moderns are able to analyse the creations as the combination of the two 'pure' binary forms (Latour 2001, pp. 34, 78).

Eco-philosophers have criticised the way the theoretical work of the moderns has made it possible for them to deny that human lives are dependent on the continuance of healthy ecosystems. To do this, as philosopher Val Plumwood described, the moderns characterise nature as lacking human attributes: mind, rationality, spirit or the outward expression of these in language and communication. With this characterisation, the moderns can deny their dependency on this subordinate or alien other. Val argued that the moderns' belief that they are dominating an inferior natural world is in denial of our very dependency on nature for survival (Plumwood 2002, pp. 10, 41, 194).

As the destruction of ecosystems on a global scale becomes increasingly apparent, human survival demands a rethink of these modern intellectual traditions. Bruno Latour observes that the moderns keep trying to continue to be modern, but in the face of this ecological devastation 'the will to be modern seems hesitant' (Latour 2001, p. 9).

As part of this intellectual rethink, governments, bureaucrats, scientists, social scientists, philosophers and others are engaging with the analysis of networks and relationships. Anthropologists are examining the cultural assumptions that form the nature/culture binary, and are building knowledge frameworks that connect people to the environment within which they live (see Strathern 1980, Ingold 2000, Ingold 1996a, Smith 2005, Rose 2004b). Ecologists are theorising towards the kinds of connected relationships and webs of life that Indigenous people have long spoken about (see Lindenmayer 2007, Manning et al. 2004). Eco-philosophers are working to reinvigorate a culture that recognises and engages with earth's life, agency and spirit (for example, Mathews 1994, Leopold 1949, Robin 2007, Main 2005).

The need for profound change in our intellectual traditions is also a part of the re-examination of water management in the Murray–Darling Basin by water bureaucracies. Roy Green, at that time the Commission's president, wrote about the value of engaging with artistic responses to today's water management challenges (Green 2002, p. viii):

> [Art] provokes and stimulates us to think about what really matters in our lives — what we should support or oppose, what is beautiful, abhorrent or special. It causes people to feel passionate, enthusiastic and determined, essential qualities if they are going to be active for the long haul.

Government policy approaches often position science, economics and management as the main factors shaping the implementation of sustainable natural resource management in the Murray–Darling Basin, but such approaches ignore the role of values. If we are going to make long-term behavioural change for benefits that in many cases will not be realised in our lifetime, it will not be primarily because cost–benefit analyses reveals that commitment to be a good economic investment. Our values define what matters and what we should do in the long run.

In critiquing the influence of modern thinking in western knowledge traditions, it is important to acknowledge the broad diversity of western traditions that are not steeped in the promise of modernity. Holistic approaches were not always so marginalised from western knowledge traditions, it is just they have recently been devalued by the knowledge of the moderns. For example, in England the Celts treated wells as holy places (Strang 2006, pp. 21–2). Kate Rigby has written about the influence of German and English romanticism on understandings of nature and place as a counter to binary thinking (2004). This diversity is not included in modern approaches. As described earlier, the moderns perceived the people who came before them as having anarchistic knowledge systems and gave them the label of 'premoderns'. In doing so, they disqualify the knowledge of the premoderns as a disposable precursor on the path to attaining modern knowledge. Indigenous peoples' knowledge falls into this category. Conversely, Bruno Latour has argued that Indigenous peoples' views of the world are not premodern but 'amodern'. Further, it is not just Indigenous peoples who are amodern, but all people: 'No one has ever been modern. Modernity has never begun. There has never been a modern world.' (Latour 2001, p. 47).

Instead of perceiving modernity as a period in time that breaks from a premodern past, modernity can be relegated to being a small part of our world histories and geographies where our lives are lived within networks and relationships. There can no longer be two distinctly different realities of the Murray River (spiritual/traditional, giving way to economic/modern); rather, the Murray River is a place where we can understand that our lives are embedded in networks and relationships. The dam builders were speaking a language of separation, but they mixed up binaries on a basin-wide scale, while also trampling valued connections and networks. The modern vision was never capable of being realised, and is instead a brutal misconstruing of relationships with the river within a long history of interconnected lives.

MODERNITY AS TIME AND SPACE: INDIGENOUS PEOPLES' EXPERIENCES IN SETTLER SOCIETIES

At the beginning of the twenty-first century, Australia is an intercultural society within which Indigenous and non-Indigenous people negotiate similarity and difference (Hinkson & Smith 2005; Merlan 2005). Through this interaction a profound syncretism has occurred (Smith 2007, p. 77). Indigenous people live closely with non-Indigenous people, and inform and translate cultural norms. Indigenous people are only 2.4 per cent of the Australian population (Taylor & Biddle 2004, p. 5), and must negotiate their own space and authority with Australian political institutions which are based on majority systems. These institutions have been intensely influenced by modern thinking. In these negotiations, Indigenous peoples have worked productively with modern thinking and have adopted modern knowledge as their own, but modern knowledges carry discriminatory categorisations for Indigenous peoples. The moderns perceive Indigenous peoples as premoderns; thus, Indigenous peoples must either conform to an idealised premodern past or risk being labelled inauthentic or even non-Indigenous.

The moderns' positioning of Aboriginal people as distant in both time and space can be illustrated by the history of anthropology. In the late nineteenth century, the discipline of anthropology was established as a science of disappearance with the focus on describing vanishing ways of life (Fabian 1991, pp. 197, 196, 200). The quest of the anthropologist was to travel into remote areas to find people whose way of life was from 'the past' and to describe their way of life 'before it is too late' (Elkin 1970). This type of anthropology is arguably a death work as it only values what Indigenous peoples can offer from the past, and assumes Indigenous peoples have no future (Rose et al. 2003, p. 30).

A related dilemma for Indigenous peoples is the moderns' theory of a hierarchy of civilisations. In this theory western culture is at the peak of civilisation while Indigenous culture is the primitive beginning of civilisation. This model relies on an assumption that before colonialism Indigenous people existed in remote and unknown isolation, without history, and without experience of the wider world to draw on to determine their own existence. Within this model any change that resulted from colonialism is understood as cultural loss for the Indigenous people. In western culture, the loss of culture is considered necessary for progress; for example, the shedding of romantic or superstitious beliefs that 'hold people

1. NARRATIVES AND THEIR RELATIONS

back'. However, the westerners can remain western, whereas Indigenous people must shed their inferior Indigenous culture. This process is also described as the model of progressive reason (Sahlins 1999). The demise of Indigenous peoples' traditional culture is predicted by the moderns as a necessary part of their move towards a higher state of civilisation, or from custom to rationality (Sahlins 1999, pp. ii–iii, xi). Anthropologist Anna Tsing summarises the moderns' narrow conceptualising for Indigenous people as having a certain type of reason that 'cannot grow' (Tsing 2005, p. 9).

The moderns associate modernity with a certain place and time — essentially the West since the Enlightenment. According to the moderns, modernity was developed in the West and then 'exported elsewhere for others to mimic', and modernity is also a rupture and break from a stable and archaic past (Mitchell 2000, p. 1; Latour 2001, p. 10). The passage of a singular historical time is an important aspect to modernity, as Timothy Mitchell has written (2000, pp. 8–9):

> The conception of historical time renders history singular by organising the multiplicity of global events into a single narrative. The narrative is structured by the progression of principle, whether it be the principle of human reason or enlightenment, technical rationality or power over nature. Even when discovered acting precociously overseas, these powers of production, technology, or reason constitute a single story of unfolding potential.

Postmoderns who critique modernity, including Jacques Derrida and Michel Foucault, still assume the existence of the West and its exterior long before European-centred dualisms had divided the world into the West and the rest (Mitchell 2000, p. 3). Timothy Mitchell concludes that modernity is a 'staging of history' not a 'stage of history', and each account of modernity and postmodernity re-enacts this staging (ibid., pp. 23, 27).

The moderns' division of time and space is a very powerful intellectual framework. Western anthropologists have learnt to suspend the nature/culture division when engaging with 'exotic' cultures and can weave these threads together in their analytical descriptions. However, when the same ethnographic work is undertaken in western cultures, nature and culture prevail as separate categories (Latour 2001, pp. 100–1). Thus the exotic cultures are judged as premodern and the western cultures as modern. This division in the ethnographic approach implies that Indigenous and non-Indigenous people are somehow living in two separate worlds. When Indigenous people in outback Australia use satellite technology to organise a hunting expedition, the moderns struggle to keep their preconceived ideas about time and space in order.

What is important to keep in mind is the influence of modernity on the authority that Indigenous people can command in settler societies. This can be illustrated with reference to the most recent recognition of Indigenous rights in Australia: native title (Strelein 2006). According to the *Native Title Act 1993* (Cth) native title applicants must prove their native title rights and interests by identifying who they are and where their country is, *and* that they continue to hold 'traditional laws and customs'. Different views by judges in the Federal and High Courts about tradition and adaptation are now determining whether native title applicants will be successful or not. These views are revealing the persistent influence of modern notions of progressive reason and hierarchical civilisations. Indeed, the government has argued that native title rights and interests cannot grow, instead they can only diminish (Strelein 2005, p. 65).

WILDERNESS AND PRIMITIVISM

Where environmental issues are concerned, Indigenous people must negotiate these models with an additional understanding of how nature/culture binaries interact with modern notions of tradition. The separation of nature and culture into opposing binaries is replicated in western land tenure systems that separate nature into definable people-free spaces: national parks, reserve lands, protected areas and suchlike. This is most well known as 'wilderness thinking'. But biodiversity needs more than just space. The dynamic flux of nature cannot be contained within land tenure and is energised by far-reaching relationships (Robin 2007, pp. 179, 175). In Australia, Eurocentric notions of wilderness that valorise nature serve to separate humans from their environment. But, rather than protecting nature, this separation has underwritten a rationalist and utilitarian approach to country (Kinnane 2002, p. 24).

Bruce Braun has studied how the conflict over the temperate rainforest in British Columbia, Canada, denied Indigenous peoples' relationships with the rainforest (Braun 2002). This conflict was dominated by two lobby groups: the foresters and the environmentalists. The foresters conceived the forest as a commodity to be managed for the nation, whereas the environmentalists were protecting a pristine wilderness that needed saving. Bruce Braun concluded that both the environmentalists' *defence* of nature and the logging advocates' *exploitation* were 'complicit in forms of erasure and abjection' (ibid., p. 2). Both the foresters and the environmentalists relied on external and universal concepts of nature, and both excluded the Nuu-chah-nulth, the local First Nation people. The foresters focused on technical expertise and scientific management, framing the issue away from social or ethical terms (ibid., p. 37). The environmentalists relied on the binary of pristine nature and destructive humanity to argue that they had the authority to speak as nature's defenders (ibid., p. 2). Yet the Nuu-chah-nulth had long lived in this part of Canada. As part of their response to the conflict over the rainforest, the lawyers for the Nuu-chah-nulth commissioned archaeological consultants to prepare a map of 'culturally modified trees' (trees that show evidence of Nuu-chah-nulth activity, ranging from felled trees to those stripped of bark) as evidence of continued use of the forest. This map of the Nuu-chah-nulth's activities in the forest overturned the theory that the forest was empty, instead showing it as a social and cultural place (ibid., p. 101).

When faced with the continuous use and occupation of the forest by the Nuu-chah-nulth the environmentalists erased the First Nation peoples' contemporary presence by collapsing them into the same category as the forest: nature. The environmentalists perceived that the Nuu-chah-nulth were living in a premodern harmony with nature which must be protected from a threatening modernity (ibid., pp. 71, 12–13). In a book of photographs produced as part of the wilderness campaign, the presence of the Nuu-chah-nulth was depicted by silent artefacts and objects (a lone totem pole), covered in moss and slowly disintegrating, sending a message of the tragedy that necessarily results from the fatal contact of modernity with a primitive people (ibid., p. 83). Bruce Braun wrote that the book assigns native culture to the premodern, likewise that the forest is 'relegated to an anachronistic space anterior to the present' (ibid., p. 86). Both the Nuu-chah-nulth and the rainforest are in a place that is timeless and separate to modern Canada. The prejudicial framework of hierarchical civilisations is also relied upon to collapse Indigenous people into nature. As primitives, the Nuu-chah-nulth are less evolved as humans, and are thus more closely associated with the plants and animals.

Complex ground is walked when Indigenous people say to governments, who are under the sway of modern knowledge, that the lands and the river and the people are one. Aboriginal people run the risk of appealing to the prejudicial framework of primitivism when they wish to express the importance of ecological relationships. They are either collapsed into the landscape or, if their identities profoundly confound this collapse, they are perceived as not really being Indigenous. This important information about how to live in the world persists in being confounded by binary thinking at a time when life-sustaining ecological ethics are really needed.

ECOLOGY AND ECONOMY

Both Indigenous people and environmentalists wishing to engage in debates that match nature and economy are challenged by the ecology/economy dualism. This dualism positions environmental issues, and by extension environmentalists, as oppositional to economic development agendas. However, the ecology/economy oppositional relationship is one of many modern dualistic assumptions now being challenged by ecological devastation. In the early 1990s the Australian Government developed 'ecologically sustainable development' as a central policy platform as part of the international policy and activist movement that is now represented by the language of sustainability (DEH 2007). The goal of the sustainability policy was given as:

> Development that improves the total quality of life, both now and in the future, in a way that maintains the ecological processes on which life depends.

This policy asserts that positive and sustainable outcomes for both ecology and economy can be simultaneously achievable goals, and at the time of publication was expressed in over 120 Australian laws (Dovers 2003, pp. 141, 144). Despite this policy step, the positioning of the ecology/economy binary as an oppositional relationship remains profoundly influential for people who are advocates of sustainability, including Indigenous people. When Indigenous people express the value of their close ecological relationships, these relationships are perceived as not having economic value. Indeed, Indigeneity is currently popular as a remedy for modernity, as part of the negative revaluation of modernity in response to global environmental devastation (Braun 2002, pp. 92–3).

The presumption that Indigenous people are against economic development has led to tension between modern thinking environmentalists and Indigenous people when both parties seek to make an alliance within an environmental debate. In these alliances, Indigenous people are sometimes represented as ecological saviours, required to have certain values that are antithetical to capitalist or market economy pursuits (Lohmann 1993). In this context, if Indigenous people appeal to conservation arguments to gain leverage in order to obtain rights to their natural resources, they have to remain free of commercial aspirations. If Indigenous people choose to use this double-edged sword, then at the same moment that they realise their rights to their natural resources they deny themselves the capacity for self-determination (Kalland 2003, p. 170). As Indigenous geographer Marcia Langton has pointed out (1995, p. 16):

1. NARRATIVES AND THEIR RELATIONS

> The ambiguous relationship of Australians with the persistent Rousseauian ideal of the savage in cultural forms, images, representations, ideas, politics and laws, has significance for how indigenous people are able to advance the restitution of their traditional territories, jurisdictions and control of resources.

The Enlightenment philosopher Jean-Jacques Rousseau, cited by Marcia Langton, is popularly associated with the concept of the noble savage, who is a person untouched by civilisation. Similar theorising was relied on by environmentalists when they collapsed the Nuu-chah-nulth identities as living in a premodern harmony with the rainforest.

It is important to consider how knowledge is formed because this has consequences for who can speak and with what authority. Modern representations — such as nature, nation, economy — carry power relationships that authorise certain types of knowledge. In Australia, Indigenous people find themselves engaging with the moderns' knowledge frameworks, and this has particular implications for their responses to ecological devastation. When Indigenous people undertake work that is supportive of ecological life, it is categorised as culture and thus separate from economy. As their culture is also categorised as premodern, Indigenous people are thus relegated to a conception space apart from modern land management.

It is only by following the threads of knowledge-formation that those people heavily influenced by modern knowledge can destabilise their presumed universal frameworks, and be open to hear what Aboriginal people have to say about the Murray River. Otherwise the moderns will continue to dismiss the statements of love, loss and connection made by the traditional owners as irrelevant to the science and economics of modern water management. With a critical take on modern knowledge, all people can also better understand how ecological devastation occurred in the Murray–Darling Basin, and be alert to government responses to ecological devastation that remain constrained by modern assumptions.

CHAPTER 2

Water management in the Murray–Darling Basin

You can't manage what you don't measure.
John Howard, former Prime Minister of Australia[1]

The Murray–Darling Basin can help us understand this theoretical analysis in terms of connections, rather than separations, because it is not a pristine wilderness that needs to be locked off for its own protection but an extended network of rivers linking communities, livelihoods and life. The Murray–Darling Basin is a large inland river basin in south-east Australia that has been transformed by government and private investment in water infrastructure to provide irrigation for the agricultural industry. This area is now known as Australia's agricultural heartland. However, extensive river degradation has threatened this productivity. The destruction of freshwater ecologies, and the far-reaching impacts of this destruction, is demanding a rethink of water management, law and policy. Today, the rivers carry water that is dramatically reduced in quality. Persistent drought has tipped a precarious system of over-allocated river water into catastrophe. This destruction has happened in a very short time; indeed, it was only in 1824 that the first European explorers sighted the Murray River.

For the traditional owners these are their ancestral lands. Before colonisation the Murray River was one of the most densely populated parts of Australia (Webb 1984, pp. 169–70). Today, Indigenous people constitute 70 000 of the over two million people who now live in the Murray–Darling Basin. This represents 3.4 per cent of the Murray–Darling Basin population and 15 per cent of the national Indigenous population (Taylor & Biddle 2004, p. 4). The traditional owners' political arguments about their unique position within this country are crowded by the many other interests of people who seek secure access to increasingly scarce, increasingly degraded and increasingly economically valuable water.

Dam builders must think of water in identifiable quantities if they are to make legible the representations necessary to engineer their water projects, but the

---

1. (Howard 2007, p. 15)

constant movement of river water keeps connecting people, places, plants and animals. The dam builders' failure to induce water to conform to their narrow conceptualising has led to widespread ecological devastation, threatening the very communities and livelihoods that the water projects were meant to support. Also threatened is their reductive view of what water *is*.

WATER AND RIVERS, MOVEMENT AND CONNECTION

Unfortunately for the moderns, it is particularly difficult to define river water as an external resource that has universally conforming properties. Water flows through towns and farms, filters through the soil, and is absorbed and released by plants and animals. People drink, swim and bathe in water. Water has moods, reflects images, and has different ages and qualities. Crucially, water is a key connecting life force because all life needs fresh water to survive.

The lands and waters of the Murray–Darling Basin provide particular challenges to a water management science based on European knowledge. Australia is an old continent that has not had much geological activity, and thus is very flat with low levels of nutrients in its soils. Large parts of the Murray–Darling Basin used to be under seawater, and salt remains a feature of these ancient soils. In this old flat country, water supplies are extremely variable and unpredictable; the country is often either in drought or flood. These variable water supplies are also incredibly small in terms of rainfall and run-off (Lindenmayer 2007, pp. 5–6).

Low rainfall means that river water is critical to the survival of life in the inland river country. The Murray and the Darling rivers, the two longest rivers in Australia, bring water from the high ranges of the east and carry it west and south through the long flat plains of the dry inland (see figure p. 28). Very little water is received from any run-off from the land out west in the semi-arid and arid inland country. It is the ability of these rivers to bring water from the east and the upstream mountains of New South Wales, Victoria and Queensland that has sustained life in the semi-arid and arid inland.

These inland rivers follow the characteristic profile of variability. They cannot be described as regular channels. River floodplains comprise up to 90 per cent of the river's path (Kingsford 2002, p. 3). The Darling is more variable than the Murray, tending to flood from the northern tropical summer rainstorms, and with longer and more frequent periods of low flow (Young et al. 2001b, p. 9). The Murray has more consistent and substantial water sources, and, before river regulation, used to have highest flows in winter and early spring. The Darling joins with the Murray in far-western New South Wales. The 1956 Great Flood of the Lower Murray occurred

The Murray–Darling Basin covers one-seventh of the continent and includes the three largest rivers in Australia: the Murray River, the Darling River and the Murrumbidgee River.

when unseasonable autumn rains in the Murray catchments and late summer rains in the north meant that the Murray and Darling rivers both flooded at the same time.

The cycles of flood and drought are understood as part of the weather pattern known as the Southern Oscillation (ENSO; see Smith 2001, p. 197, Hennessy et al. 2004). Understanding this influence of weather pattern is part of our growing appreciation of the earth as an interconnected ecology. ENSO describes a relationship between ocean water temperatures and rainfall: when water temperatures rise in the Pacific Ocean, rainfall decreases in certain parts of the world, including the Murray–Darling Basin. When ocean water temperatures cool, there may be above average rainfalls and flood events.

The inland river variability is central to the robust productivity of complex ecologies. Rather than variability being an irregularity from what is assumed normal for a river, it is a key to life. Plants and animals have adapted to live and thrive with variability. Many plants regulate water loss through their leaves, recover rapidly after stress, and have developed extensive high-density root systems to capture water. Numbers of small mammals, frogs and reptiles can increase rapidly after rain. Nomadic birds travel large distances to where the rains fall. The female red kangaroo can delay the development of an embryo until rainfall brings grass and water to the inland plains (Lindenmayer 2007, p. 5). The iconic river red gum, which lives next to the rivers, sheds leaves during drought and regenerates best after floods (Young 2001, p. 190–1). Golden and silver perch spawn when the floodwaters come (Wahlquist 2005). Thus, ecological dynamism thrives on the complexity of the flooding regimes, and this complexity supports stability and productivity. The floodplains, wetlands, freshwater and saline lakes, anabranches, billabongs, lagoons, overflows, swamps and waterholes are places of extraordinary biodiversity. The more robust the connectivity in the landscape, the more resilient animals and plants are to large disruptive events such as bushfires or extended droughts.

The physical attributes of water are critical for all organisms within webs of life, but water is so much more than even this vast life-sustaining presence. Water encapsulates landscapes, insinuates borders, provides images that inspire artists, and comes in a variety of colours, shapes, smells, qualities and sounds. Torrents generate a loud energy, whilst still waters create places of peace and solitude (Toussaint 2006, p. ix). These are sensory experiences: felt, heard, seen, smelt and tasted. Water is encoded with meanings from intimate interactions involving 'ingestion and expulsion, contact and immersion' (Strang 2006, p. 5). Far from being external, water travels through our bodies, and is part of our form. A 70 kilogram human has 41 litres of water sluicing within (Boulton & Brock 1999, p. 15). With two million people living in the Murray–Darling Basin, that's 82 megalitres of water running through the human population.

Geographer Leah Gibbs researched the experiences of people living in the Lake Eyre Basin in central Australia. Leah observed that 'water is not just water'; rather, there is rain, river, and bore water, and within that people have more nuanced appreciations. River water is different when in the channel compared with flowing over the land; rainwater is described as soaking, light, steady or follow-up rain; and bore water at South Galway pastoral station is good drinking water, while bore water at Mungerannie pastoral station has a strong sulphurous smell (Gibbs 2006, p. 77). Understanding water also requires understanding how it relates to place.

The Lake Eyre Basin is a large network of rivers that are usually in drought, but with the flood waters a spectacular diversity of life thrives. Because of this, floods are celebrated. Station owner Sharon Oldfield spoke to Leah about the floods: 'You go out and you can see it and it just lifts your heart' (ibid., p. 81).

The landscapes forged by the rivers in the Murray–Darling Basin are an iconic part of Australian imagery. Don Blackmore, former chief executive of the Murray–Darling Basin Commission, described these inland rivers as the 'quintessential image in our mind when we think of Australia' (Blackmore 2002, p. 1). The rivers are known for their slow meandering twists and turns and snake-like folds. The rivers have played a large part in recent history. They provided early colonial explorers with routes into arid Australia, and sustenance for these journeys. Today they are celebrated as beautiful places for swimming, fishing, water skiing, and boating. The 'Majestic Murray', stretching 2530 kilometres from Alps to ocean, is the longest river in Australia. The mighty river red gum, reaching up to 45 metres in height and 1000 years in age, is the 'greatest symbol of the greatest river' (Larkins & Parish 1982, p. 61). The Barmah–Millewa forest, on the Murray in the riverine plains, is the largest river red gum forest in Australia. Australia's largest freshwater fish, the Murray cod, can weigh in excess of 100 kilograms and grow up to 2 metres in length, and is a cultural icon for both Aboriginal and non-Aboriginal people (Sinclair 2001, pp. 120–40; Humphries et al. 1999, p. 130). The national symbolism of this place is such that 'To understand the [Murray] river is to come to terms with Australia' (Larkins & Parish 1982, p. 7).

Aboriginal people tell Dreaming stories that embed the inland rivers as places of energetic spiritual action by the ancestors. Rather than just one story, each language group has its own stories about how their country was created. One of the most well-known Dreaming stories of the Murray River is that of *pondee*, the Murray cod. The Ngarrindjeri relate how this giant *pondee* was chased down the Murray River, from the junction with the Darling River, by the ancestral being Ngurunderi who was trying to spear the fish. The *pondee* thrashed through what was a small stream, widening it by the movement of its strong tail and thus creating the Murray River in what is now known as South Australia. When the *pondee* was caught it was cut up and the pieces of the *pondee* became different fresh and saltwater fish species to sustain the Ngarrindjeri people. Further upstream the Yorta Yorta people, whose country includes the Barmah–Millewa forest, tell of Baiame's creation of Dhungala (the Murray River). Baiame sent a giant snake to follow his wife as she travelled from the mountains to the sea. The path of the giant snake made curves creating the riverbed; this was later filled with rainwater to form Dhungala. Such stories tie people to their distinctive part of the river in a potent, spiritual way.

The moderns must not accept these profound and diverse meanings of water — the feelings, sensory experiences, and Dreaming stories — if they wish to separate and describe water as a definable resource with stable knowable properties. To accept these meanings is to bring many other considerations into their decision-making context. But despite the moderns' efforts, the water persists in emphasising these relationships and connections, and its intimacy with life. The people who live close to rivers keep responding to these meanings. Instead of an abstract resource, they articulate water as a key connecting life force. These relationships with water are not weak or distant networks, but are experienced as strong life-giving connections.

WATER MANAGEMENT IN THE MURRAY–DARLING BASIN

The Murray River has been perceived by governments and others as central to the economic potential of the nation. This has involved an embracing of the moderns' conceptualisations of nature, economy and nation — and water. These representations have enabled governments to embark on ambitious water projects with power and authority.

In the late eighteenth century, colonialists recognised that the Murray River would be a liquid lifeline for agriculture in the semi-arid and arid inland, extending colonial settlement beyond the fertile eastern coastal strip (Powell 1989, p. 137). In the 1940s and the 1950s, governments and private industry popularised the Murray River as a powerful unlimited resource for the production of agricultural crops. With economic production as the goal, water was described as liquid gold (Sinclair 2001, pp. 76–9). The 'magic touch of water' could be harnessed to transform 'useless country' (Murray Valley Annual 1951, cited in Sinclair 2001, p. 78). Today, 90 per cent of the water consumed in the Murray–Darling Basin is used to irrigate agricultural lands (Ball et al. 2001, p. 21). Irrigation effectively diverts water into new networks, expanding the system of waterways from ephemeral creeks to regulated channels next to irrigated fields (see figure following). The Murray–Darling Basin accounts for 41 per cent of the value of agriculture produced nationally, giving the region the title of Australia's 'food basket' (MDBMC 2001, p. 1).

However, the eighteenth-century colonialists were thwarted by the highly variable flow of the inland rivers, which was at odds with the methods of European agriculture. The colonialists suffered valuable crop and stock losses as they lacked knowledge of the variable weather patterns. Irrigation schemes to 'drought proof' agriculture became an important political issue during the 1870s and 1880s in Victoria (Smith 2001, pp. 200, 203–4). In 1886, Victoria became the first colony to pass legislation vesting far-reaching control of water in the colonial authority

The Moira Channel — an example of the new journey that water makes in the Murray–Darling Basin.

(Connell 2005a, pp. 83–4). When the floods did come, many towns had to be moved, including the Murray River town of Moama (Smith 2001, p. 229). In June 1852 the Murrumbidgee River flooded Gundagai, killing over 80 people. The local Wiradjuri people had repeatedly warned that the Murrumbidgee was capable of huge floods, but the new settlers could not conceive the scale of that warning. Afterwards, Gundagai was rebuilt on higher ground.

Controlling and allocating inland river water was part of the negotiations when the colonies came together to form the federation of Australia in 1901. The southern bank of the Murray River forms the border between Victoria and New South Wales, before the river flows into South Australia. In these negotiations, New South Wales sought exclusive rights to the Murray's waters and all tributaries within that state; Victoria sought a share of water diversions from the Murray; and South Australia was concerned about maintaining navigational access along the Murray. In 1914 the Murray Waters Agreement was signed and then ratified in 1915, and the River Murray Commission was established. South Australia was guaranteed a certain amount of river flow, and New South Wales and Victoria were to equally share the flow of the Murray at Albury (Connell 2007, p. 95).

With this agreement it became possible for the states to go ahead with the construction of weirs, locks and dams for the storage and delivery of water. State ownership and control of water was consolidated in the Constitution, with the power to intervene reserved for the Commonwealth under section 100. The states were able to order, control, licence, allocate and charge for the water. Today, individuals and companies apply to state governments for water permits, licences, allocations or entitlements, which are issued as use rights rather than ownership. Use rights confer the authority to take water from a water source (Productivity Commission 2003, p. 49). In recent times, control and allocation systems have been increasingly extended to groundwater, with growing recognition that all water sources are connected. The water in aquifers and rivers is interdependent; depending on the height of the watertable, groundwater either recharges water into rivers or is fed from river water.

The riverine plains in southern New South Wales and northern Victoria became the centre of irrigation in the Murray–Darling Basin. Here, weirs have raised the height of water so that it can move by gravity to agricultural lands along canals and channels. Large irrigation schemes have been developed by government and private industry to entice smaller settlers. One of the largest is the Murrumbidgee Irrigation Area. In 1927 more than 24 000 hectares was under irrigation, and this increased to 91 000 hectares by 1971 (Kingsford 2003, p. 75). Griffith, Leeton and Yanco were established as part of the Murrumbidgee Irrigation Area and Districts. Tens of thousands of people have been attracted to the irrigated lands to grow rice, horticultural crops, cattle and poultry for eggs (Crabb 1997, pp. 104). By the mid-1970s, practically all of the water in the Murrumbidgee had been allocated to irrigators (Sinclair 2001, p. 107).

After 180 years of colonisation along the Murray, Aboriginal and non-Aboriginal people live in the same sociocultural space. Here, pastoralists breed livestock and harvest cereals, labourers shear sheep, pick fruit and vegetables, and tourists come to fish and travel on the old river boats. The traditional owners participate as pastoralists, labourers and tourists, and they also carry their inherited responsibilities to country. They identify their traditional country enmeshed with this rural landscape. Country is now fragmented and marked out by freehold and pastoral leaseholds, the squares of irrigated fields, the long lines of the roads and railways, and the life of the rural towns and communities.

Before the introduction of European agriculture, Aboriginal people valued wetlands for their productivity, including food. Near Wagga Wagga, certain wetlands were protected as sanctuaries for swans, ducks and other water birds so that these birds could breed and replenish their populations (Gilmore 1963, p. 118; see also Main 2005, p. 25). Just south of the Murray–Darling Basin in Victoria, Gunditjmara people created vast fish and eel traps that operated when floodwaters spread across the land (Lourandos 1987). But the governments and private industry drained wetlands and replaced them with deeper, more permanent dams, creating both land for pasture and a water supply (Boulton & Brock 1999, pp. 170–1). The governments' implementation of modern water management practices have brutalised the river ecologies that were previously enjoyed by the people living next to the inland rivers. This is the result of a perspective that understands wetlands as part of 'useless' country.

ECOLOGICAL DISORDER

The last 60 years has seen the most dramatic increases in the consumptive use of water and alterations to the flow of rivers in the Murray–Darling Basin. Advances in engineering made it possible for the dam builders to construct immense water projects. Storage capacity grew rapidly in the mid-1950s, with concomitant increases in water diversions for irrigation. Weirs and locks across the Murray River have created a series of 'stepped pools'. Throughout the Basin, canals, channels, weirs and locks strap the rivers into a vast and complex engineering system. The variability of the inland rivers has been regulated to create a controlled water supply and to reduce the impact of floods and droughts. This regulation has restricted natural variability and the flooding patterns on which biodiversity has depended.

The impact of governments' faith in modern water management on the river ecologies has been staggering. This 'magic touch' is responsible for enormous ecological devastation. The frequency, duration, and timing of floods has been abruptly changed: instead of droughts and floods, upstream storages now deliver a slow flow of predictable water down the Murray. Floodplains throughout the Basin are bone dry. The Chowilla Floodplain in South Australia has had flooding events halved by upstream dams and diversions (Kingsford 2000, p. 115). Floodplains mostly receive water from high river flows that flow out onto country from the river, rather than local run-off, and thus are severely affected by the reductions in the size of river flows (Reid & Brooks 2000, p. 481). The flow at the Murray Mouth is now on average 25 per cent of what it was before the large storages were built (DWLBC n.d., p. 2). In other places, the constant presence of water is problematic. Ephemeral wetlands that are fed a continual flow of water are never allowed to dry out, and ephemeral lakes have been drowned as they become permanent water storages.

2. WATER MANAGEMENT IN THE MURRAY–DARLING BASIN

The seasonal operation of the Murray River System is problematic because water flows are managed according to the priorities of agricultural growing seasons and not river ecologies. As water is needed for summer crops, the Murray's flow on the riverine plains is now, profoundly, the reverse of seasonal patterns (see figure below). In the riverine plains, water has to be released from the Hume Dam four days ahead of demand to allow for time to travel to irrigators. The Murray River is used to channel this water to downstream irrigators. However, if summer rains fall then the irrigators may cancel their orders after the release. These 'rain rejection' flows inundate the Barmah–Millewa forest and wetlands at a time of year when the wetland is drying out (Chong & Ladson 2003, pp. 162–3). The lower areas are too wet because of the many small floods, but without the large floods the higher areas of the Barmah forest are too dry (ibid., p. 163).

Water quality has suffered from this water regulation, as well as from land clearance for agriculture. Changes in land use have increased erosion and sediment in the rivers, altering the organic matter entering rivers and introducing chemicals and high levels of phosphorus and nitrogen (Young et al. 2001a, p. 51; Young et al. 2001b, pp. 4–5). Sediment loads in the Murray–Darling Basin are now estimated

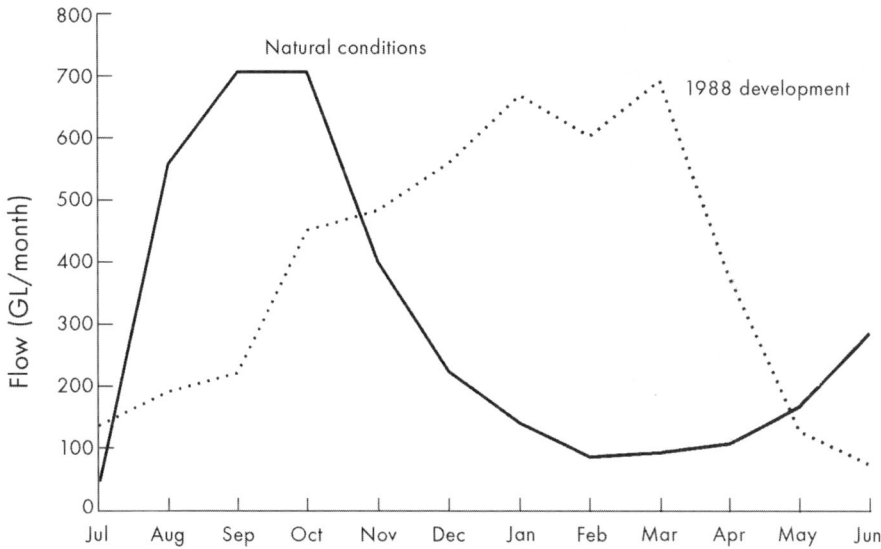

Modelled median monthly flow volumes (gigalitres) on the Murray River at Albury under 'natural' and '1988 development' conditions (Young and Hillman, 2001). © Murray–Darling Basin Authority.

to be 41 times higher than before river regulation. River water quality has also been changed by large reservoirs. Cold water released from the bottom of these reservoirs suppresses water temperatures for up to 300 kilometres downstream (Gehrke et al. 2003, p. 5). In the summer of 1991–92, over 1000 kilometres of the Darling River was affected by deadly blue-green algae (cyanobacteria), forcing the New South Wales Government to declare a state of emergency. The increasing salt content of the rivers and the land together form one of the most dramatic ecological challenges. The inland flat country, with relatively low river flows, is unable to flush salts and sediments out to sea (Connell 2007, p. 18).

The impact of these changes is evident in the loss of river life. Simplistic approaches to complex ecologies have undermined the dynamism and diversity that previously supported ecological productivity. Murray crayfish are now locally extinct in the Lower Murray, and snails and mussels have suffered huge declines in numbers (Gehrke et al. 2003, p. 6). 'Rain rejection' flows in the Barmah–Millewa forest cause tree deaths, the disruption of nutrient cycles and the loss of the Moira grass plains (Chong & Ladson 2003, pp. 162–3, 165). For group-nesting waterbirds who have long relied on this place — such as the magpie goose, black swan, glossy ibis, and certain grebes, herons, and cormorants among many others — successful breeding episodes have been reduced by 80 per cent, largely because breeding cues are linked to the magnitude, duration and frequency of floods (Leslie 2001, pp. 23, 33–4). Because of the relationship between lifespan and the frequency of breeding cycles, much of the current generation of birds may not be replaced.

Native fish have been dramatically affected. Current fish populations are estimated at 10 per cent of pre-European settlement levels, with most of the decline occurring in the last 50 years (MDBMC 2003b, p. 6). Sediment loads bury fish breeding sites, and cold water temperatures disrupt fish breeding cycles. Major physical barriers have led to the local extinction of migratory fish upstream (Gehrke et al. 2003, p. 6). When the Dartmouth Dam was constructed in 1980, Murray cod, Macquarie perch and trout perch all became locally extinct in the Mitta Mitta River (MDBMC 2003b, p. 6). Overall, eight of the 35 native fish in the Basin are threatened with extinction, with two species listed as critically endangered. The introduced carp now accounts for 90 per cent of the fish biomass in some river stretches in the Murray–Darling Basin, after a particularly invasive strain of the carp was introduced in Victoria in 1961–62 (CCCG 2000, pp. 1, 3).

Following this long list of interventions and modifications, when the weather goes into a drought cycle the river ecologies — already starved of water — are pushed to the brink of survival. Complex ecological relationships have been sent deeply awry. In the Barmah–Millewa forest, when there is a rare flood the prolonged build-up of forest leaf litter leaches tannin into the river, de-oxygenating the water. When a flood came through in 2000, thousands of Murray crayfish and yabbies were forced to crawl out of the water in search of oxygen, only to die on the riverbanks. In their desperation, they were even seen climbing trees. The story of the Murray and Darling rivers is a story of 'ecological disorder' (I have borrowed this term from Main 2005). Complexity and connectivity have decreased, and the life that that connectivity sustained is heading towards death.

The future of people living by the river has also been put at risk by the disregard shown for the web of social and ecological relationships. Most commercial native fisheries have closed and recreational anglers report that only 4.4 per cent of their catch is native fish species (MDBMC 2003b, pp. 1, 6). The commercial fish catch of Murray cod has reduced from 74 tonnes per year in the late 1940s to 9.5 tonnes per year in the early 1990s (MDBMC 2003b, p.12). Businesses that rely on secure fresh water entitlements are facing downsizing or closure. With ongoing drought some towns have had to truck in water to meet domestic needs. River water is now so polluted that swimming carries serious risks of infection. Holiday houses and tourism business have been left high and dry by low water levels in water storages. Tourist magazines continue to promote the Murray River as a place of romance and beauty, but they are finding it harder to continue the deception.

It is the farmers whose threatened viability has made the most impact in the newspapers and parliaments. Farming the land is still central to Australian images of national development success, and continues to be perceived as a business of national importance (Robin 2007, pp. 208–9). But the tight relationship between farming success and national identity is deeply challenged by the failure of the large water projects to support the agricultural industry. In New South Wales, water allocations are being made on a month-to-month basis. Farmers who need secure water supplies to sustain longer-term crops, such as grape vines and fruit trees, are hard hit. Irrigators in South Australia have been struggling to keep trees alive with just 13 per cent of their water entitlement (Harmsen 2007).

RESPONDING TO THE ECOLOGICAL CRISIS

The devastation has been so severe and far reaching that it is now the 'crisis' of the Murray River that has captured the imagination of the nation. Drought conditions have dominated the first decade of the twenty-first century. The water issues of the Murray–Darling Basin have remained front page news. On television, weather presenters forecasting the likelihood of rain do so with humility, cognisant of farming communities living with extended drought.

Before this current drought, governments had reassessed water management to ensure the future quality and quantity of water supply. From the 1950s onwards, salinity had been growing as a community concern, particularly in South Australia, at a time when ecological awareness was also on the rise (Connell 2005b, p. 148; Sinclair 2001, p. 223; Connell 2007, p. 105). When droughts periodically came, salinity levels dramatically worsened, prompting community pressure for change. In the 1980s, reassessment of the major water institutions gathered momentum. Of profound significance and symbolism, the 1981 river flow was so low that the Murray Mouth closed for the first time in oral and recorded history. With South Australia pushing strongly for change, the institutional structure we now know as the Murray–Darling Basin Initiative was created in the 1980s as a government–community partnership to respond to water quality issues (Connell 2007, p. 9; Blackmore 2002, p. 2). Its establishment marked the creation of the Murray–Darling Basin as an identified administration area for coordinated water management.

The Murray–Darling Basin Initiative is led by the Murray–Darling Basin Ministerial Council (the Ministerial Council), which itself comprises ministers from each of the six interested jurisdictions: the Commonwealth, the Australian Capital Territory, New South Wales, Queensland, South Australia and Victoria. The Murray–Darling Basin Commission (the Commission) provides advice to the Ministerial Council and manages the water-sharing arrangements. The third institution involved in the Initiative is the Community Advisory Council (CAC). The CAC directly advises the Ministerial Council on community issues, and assists the Ministerial Council in engaging with the communities in the Basin (Morgan et al. 2006). (These arrangements were overhauled in late 2008 to create a single agency responsible for the management of water resources in the Murray–Darling Basin (*Water Act 2007* (Cwlth)). The responsibilities of the Commission were assumed by the Murray–Darling Basin Authority, which now implements the decisions of the new Ministerial Council and Basin Officials Committee. The CAC was replaced by the Basin Community Committee.)

Sustainability has become a central concern for the Murray–Darling Basin Initiative. Sustainability is often described as 'the triple bottom line': action that has economic, social and environmental benefits. The 1992 Murray–Darling Basin Agreement, which underpins the Murray–Darling Basin Initiative, has as its purpose (under the *Murray–Darling Basin Act 1993* (Cwlth)):

> to promote and co-ordinate effective planning and management for the equitable, efficient and sustainable use of the water, land and other environmental resources of the Murray–Darling Basin.

Within this overhaul of river management, specific policies have been developed to improve water quality. In 1995, the Ministerial Council introduced a cap to limit river water extractions at 1993–94 levels (Yencken & Connell 2002, p. 71). This policy signalled a clear 'line in the sand' to setting some limits on development agendas. Environmental water allocations, or 'environmental flows', were also introduced. An environmental flow is a return of water to a river as close as possible to the natural hydrological rhythms of the river (Thoms et al. 2004, p. 353). Environmental flows mark a recognition that water is needed to support the health of freshwater ecologies, and are an explicit reversal of purely extractive models of river management.

Disappointingly, environmental flows and the cap have not led to sufficient positive change to improve river health. The cap assumes that 1993–94 levels of water extraction are reasonable for river restoration (Connell 2005b, p. 179) and does not limit extraction from groundwater reserves, although a 2000 review of the cap recommended it do so (MDBC 2000). The cap is also challenged by the logistics involved in monitoring water consumption and the Queensland Government's ongoing reluctance to sign. The main stumbling block to implementing environmental flows has been a lack of political will to allocate water for the environment. This reflects the perception among water entitlement holders that the environment is a competitor for water rather than fundamental to sustaining freshwater ecologies (Sinclair 2001, p. 233).

With the impacts and effects of river degradation being felt broadly in rural communities, water law and policy has become more inclusive. These changes have been in response to public criticism that important water decisions were being made by small groups of technocrats (Connell 2007, pp. 90–1). In 2001 the Ministerial Council released the integrated catchment management policy, which recognised all people as stakeholders in the Basin's health and sought to balance social, economic and environmental considerations (MDBMC 2001; Blackmore 2002, p. 4). Catchment by catchment, communities and governments were to agree on targets for water quality (including salinity and nutrients), water sharing (between consumptive uses and environmental flows), riverine ecosystem health, and terrestrial biodiversity. The policy focused on improving the decision-making process by moving it to the regional level.

In 2002 the Ministerial Council established 'The Living Murray', a large program to return water to the Murray and reduce water consumption. The program's aim was to achieve a 'healthy working river', which the Ministerial Council defined as 'one that is managed to provide a compromise, agreed to by the community, between the condition of the river and the level of human use' (MDBMC 2002,

p. 47). Ironically, The Living Murray has starkly revealed the ongoing lack of political support for environmental flows. Commissioned research by an expert reference panel advised that at least 1900 gigalitres of water needed to be returned to the Murray for a 'moderate probability' of achieving a healthy working river (a gigalitre is one billion litres), while 4000 gigalitres needed to be returned for a 'high probability' of achieving this (Jones et al. 2002, p. 27). However, the community consultation process for The Living Murray discussed only three reference points for the return of water: 350 gigalitres, 750 gigalitres or 1500 gigalitres (MDBMC 2002, pp. 2–3). These low targets reflect how the pressure for a 'business as usual' philosophy undermines the expressed need for immense change (Connell 2007, p. 37).

After all the advice and consultation, The Living Murray targetted the return of only 500 gigalitres to the river, to be achieved over a five year period (2004–09) (MDBMC 2005a, pp. 6–7). This amount is to be found through water recovery measures, putting off the politically thorny issues that come with addressing the over-allocation of consumptive water entitlements (MDBMC 2005b, pp. 7–8). In respect to government approaches to water management in the Murray–Darling Basin, political scientist Daniel Connell has argued that in many cases 'indefinite decline is an unspoken premise', and that the pragmatism of the achievable goals of The Living Murray is ultimately 'a philosophy of despair'(Connell 2005b, pp. 206, 285).

During this period of policy development, the weather worked against the irrigators and the water bureaucrats. In 2000, the Murray–Darling Basin entered what was to be the beginning of a severe drought. Water levels started dropping in the large water storages. Analysing the dramatic falls in farm income and agricultural export dollars, economist Saul Eslake described 2002 as the worst drought since written records were kept (Anon 2003). In 2003, the Murray Mouth threatened to close for a second time. Since then, the Murray Mouth has only been kept open by two large dredges, excavating sand around the clock, costing the Murray–Darling Basin Commission and the states of Victoria, New South Wales and South Australia many millions of dollars (DWLBC n.d., pp. 3–4). With the combination of all this attention and yet further declines in river health, the water crisis is being documented but not averted.

The water crisis is a national priority beyond the Murray River. Other agricultural regions across Australia are facing similar challenges with water quality and quantity. In 2004, the Council of Australian Governments released its *Intergovernmental Agreement on a National Water Initiative* (the NWI), and established the National Water Commission. The NWI recognised and acknowledged that water was over-allocated (for example, COAG 2004, Schedule A, p. 24). Arguably, recognising over-allocation should have led to the creation of policy to reduce water allocations, but the NWI is ambitiously written as though policy recommendations for better water allocation planning would achieve socially and economically beneficial outcomes that were also environmentally sustainable.

The NWI is more a broad statement of intention than a practical guide to implementing a better water management regime. Implementation of the NWI rests with the states, and Daniel Connell is doubtful whether the NWI will have enough momentum to overcome current approaches which almost completely ignore the dependency of life on the health of the rivers (Connell 2005b, pp. 77–8). By 2007 (three years after the release of the NWI), not one government in the country had seriously tried to implement the NWI; environmental sustainability had been too much of a challenge (Connell 2007).

Market mechanisms are also part of the policy mix to improve the efficiency of water use (water can now be bought and sold online; see for example National Waterexchange 2007). Water trading between irrigators has been in operation in limited areas on the middle and lower reaches of the Murray since the late 1980s. The NWI aims to expand this to a larger geographic area in order to reduce the competitiveness of inefficient irrigators and direct water to the most efficient users. Water is to be bought by those irrigators whose use of water delivers the best economic return; thus, market mechanisms will introduce considerations about water scarcity into the decisions made by irrigators as scarcity translates into high water prices. Indeed, water prices have rocketed from between $100 to $200 per megalitre (1 megalitre equals 1 million litres) for a 'temporary' annual lease of water to $1000 per megalitre (Wahlquist 2007, p. 1).

Significantly, these market mechanisms do not attempt to make any water savings; that is, reduce the current problem of over-allocation to existing entitlement holders. Trading also raises the question of whether the benefits will be outweighed by the transaction costs of moving water between different catchment areas (Connell 2007, p. 204). The NWI intends that water trading will improve environmental sustainability; however, the complexity of the regulatory environment could result in the reverse (ibid., p.205). Water trading has required the separation of water title from land titles, which creates a profound problem. By separating water from land there is the possibility that towns, infrastructure and irrigators may become stranded without water (Productivity Commission 2006, p. 91). The Commonwealth government has assured rural communities and water-dependent businesses that the trading environment will be tightly regulated to minimise such negative impacts.

Despite all these policy initiatives the drought continued and the rains did not come. This led John Howard, the Prime Minister of the day, to warn Murray–Darling Basin farmers that there may be no water allocations for either irrigation or environmental flows at the opening of the 2007–08 'irrigation season' (Coorey 2007). Forced to specifically address over-allocation, the Murray–Darling Basin Commission introduced a voluntary buyback scheme for water entitlement holders, reaching the target of 20 gigalitres four weeks into the 11-week buyback period (MDBC 2007c).

Throughout this period of responding to ever-worsening river health, government water policy shifted to make more allowance for the needs of rivers. But water management continues to be dominated by technological approaches to improve the extraction and delivery of a water resource. The rhetoric of support for river health is evident, but the political will to challenge the assumed terms of the relationships — between people and water, and between water entitlement holders and the river — is lacking. Instead, action to address the water crisis is focused on 'gold-plating' the system; that is, a continued focus on improving the technology (Mike Young quoted on Mares 2007). The governments believe they are involved in a very large plumbing exercise.

FAT DUCKS, FAT CATTLE: FAT CHANCE

The ecological destruction in the Murray–Darling Basin has happened in a very short time, but it will take a much longer time to recover. Organisms such as invertebrates and small fish may respond quickly to the return of water, even in a matter of months. Species such as yabbies, birds, turtles, platypuses, red gums and large fish, however, will take years or decades to recover (Jones et al. 2002, pp. 28, 32). The recovery of saline lands could take centuries. Nearly all parties engaging with this issue pragmatically assume that the rivers will never return to the way they were before the large water storages were built.

Despite the importance of water and its management, the development of water policy has occurred sporadically in reaction to water crises, with advances in water law and policy coinciding with periods of drought (Botterill 2003). This ad hoc approach to water policy fails to engage with the 'precautionary principle', which is a central part of sustainability. The precautionary principle was devised by policy-makers and ecologists to argue that (DEWHA 1992, part 1):

> where there are threats of serious or irreversible environmental damage, lack of full scientific certainty should not be used as a reason for postponing measures to prevent environmental degradation.

2. WATER MANAGEMENT IN THE MURRAY–DARLING BASIN

Water policy initiatives that only occur in response to drought, do not create a context for long-term precautionary planning to prevent environmental degradation in the first place.

The precautionary principle recognises that science has limited predictive capacity, whereas the impacts of development can have effects that are difficult or impossible to overcome. We are seeing this played out in the devastation of the inland river country. The dam builders and the people allocating water entitlements thought they were the ones operating the water management system. They were 'running the river'. They did not give enough consideration to the deplorable consequences of their actions.

Whilst the thinking behind sustainability relies on modern categories of economy, society and environment, it is a policy approach that moves away from the separation of nature from humans and human activities, or 'wilderness thinking' as discussed in Chapter 1, to reintegrate nature and development. The intention is to find approaches that will create 'win–win' outcomes for each category. Yet the governments that implement sustainability laws and policies remain highly influenced by oppositional dualisms, including ecology/economy. Thus, underlying much of the different governments' reluctance to reduce water entitlements across the board is the assumption that there is a necessary trade-off, or compromise, between the ecological health of the rivers and the capacity of the rivers to be 'working' rivers. According to this philosophy, river health is the unfortunate sacrifice we have to make for food production, export earnings and the life of country towns. The river has to work for us, work that we define as an economic set of objectives.

The assumption by governments that the environment is necessarily traded for economic growth is exemplified by the way the technocrats treated Psyche Bend Lagoon. This wetland, next to the Murray River in northern Victoria, used to be fed with run-off irrigation water. However, in the 1990s it was decided that the run-off would instead go straight into the Murray River in order to help reduce salinity levels. Significantly, downstream citrus orchards are seriously affected by salt content in the river water, and measuring salt levels in the Murray River is an important indicator of river health for the Murray–Darling Basin Commission. After this change in management the only water seeping into Psyche Bend Lagoon was hypersaline groundwater. In 12 months the salt levels increased by 20 times and the wetland became saltier than seawater (see colour plate between pp. 112–13). Psyche Bend Lagoon was effectively sacrificed for economic benefits downstream (Gell 2007, pp. 27–8). But Psyche Bend Lagoon is not unconnected with the river ecologies and economies that surround it; indeed, water — whether in the river or under the ground — is characterised by connectivity. If there was a flood today the toxic mix which is now Psyche Bend Lagoon would be washed out into the adjacent Kings Billabong and the Murray River, devastating agricultural crops and country. This case exemplifies how, instead of working in a way that is cognisant of ecological relationships, the moderns continue to organise water through separation and containment. The result of this trade-off thinking is to create a situation that threatens both economy and ecology.

In the Macquarie Marshes in central New South Wales the slogan 'fat ducks mean fat cattle' is used by both conservationists and graziers to highlight the role of healthy wetlands in sustaining the lives of ducks and cows. This slogan values the productivity of wetlands for both ecology and economy. Since Burrendong Dam was built upstream in 1967, the floods that used to sustain the marshes have been diverted for irrigation crops such as cotton. Irrigators and graziers dispute whether it is water extraction or over-grazing that is the problem with the marshes, but the fact that there is a problem is undeniable (hence the subtitle of this section, borrowed from Lewis 2006, p. 31). Bird numbers have plummeted from hundreds of thousands to a handful. Without water there have been no successful bird breeding events since 2000 (Anon 2007a). Unfortunately, and criminally, when environmental water was allocated to the marshes in 2006 this water was illegally diverted for private use (ibid.). Wiradjuri Elder Tony Peachey spoke to me about how many people who run water-based industries narrowly perceive the rivers as a resource to be exploited:

> Sooner or later the farmers have got to realise that the damage they are doing to the river is going to affect everyone. But I don't know how you get the message through to them, unless you hit them on the head with a hammer. I don't know what it is, just a mentality. Their main thing is, they see a big flush coming down the river, their first thought is, 'How much can I drain out?' Not, 'Hey that's neat, I haven't seen that for a while', and go and sit there and watch it.

What Peachey, as he is known, calls 'the mentality', and what I have been calling modern thinking, is what allows the abuse to keep going. In this relationship with the rivers, economic issues are still placed ahead of biodiversity. Because of this, the practice of sustainability is much more difficult than the rhetoric (Robin 2007, p. 187). As fifth-generation commercial fisherman Henry Jones remarked, the politicians are 'talking about the environment, but they are still dreaming about more production' (Jones 2002, p. 93).

The water crisis is demanding another rethink of how we manage water, but addressing the ecological crisis is challenged by the hangover of modern thinking.

Water trading keeps the focus on water as a commodity. It will introduce more-efficient uses of water in accordance with economic values — such as crops that are more profitable. But water trading continues the focus on production and does not allow the conceptual depth needed to grasp the diverse, intimate and profound values of water. Further, people lobbying governments to protect the rivers feel they have to appeal to the priority placed on economic values by governments.

There is now a struggle going on between the authority of modern approaches and the importance of the relationships that people know and hold with their local freshwater ecologies. This struggle is happening across Australia. The devaluing and subsequent deterioration of freshwater ecologies is the new universal experience. Lake Eyre Basin pastoralist Sharon Oldfield spoke to Leah Gibbs about her frustration in communicating the value of river water to government (Gibbs 2006, p. 78):

> Government policy can't cope with things that aren't tangible. I mean, how do you write government policy about something somebody feels? How do you do that? And then take it to cabinet and want funding for it.

Angus Emmott, owner and manager of Noonbah Station, also in the Lake Eyre Basin, spoke to Leah about valuing ephemeral wetlands (ibid.):

> The intrinsic values are very hard, because you can't actually put a dollar figure on them, but they're so crucial … We know they're valuable, but it's very hard to put dollar figures on them.

The states have misjudged many important matters in their efforts to harness the productive forces of the inland rivers. They unwisely believed in their capacity to bring order to ecological relationships. Order turned out to produce disorder. They did not account sufficiently for the influence of other actors in the Murray–Darling Basin, and their approach did not carry enough respect for human dependence on freshwater ecologies for survival.

When the colonies, and later the states, assumed legislative power over water, they misjudged their capacity to determine water flows. Older forms of English common law had understood a different relationship between people and water. Water was classified as something over which no person could have absolute property rights. Along with fire, light, air, and wild animals, there could only be 'qualified property' rights in water as water, fire, light, air and wild animals in law had a 'vague and fugitive nature' (Blackstone 1979).

With the persistence of the moderns' belief that the world is a passive stage for human action, water policy-makers keep planning for the control of water even when that water is no longer present. The 2007 buyback by the Murray–Darling Basin Commission of 20 gigalitres means that the Commission has reacquired entitlements for water that does not always exist, and thus the scheme will result in 'little water' being returned to the rivers (MDBC 2007c). Likewise, water-trading schemes may be planned and legislated, but they cannot operate without water. In these Kafkaesque undertakings, the lack of water is called 'dry inflow', and the Commission puts together Dry Inflow Contingency Plans ready for the 'irrigation year' (MDBC 2007b).

The evolution of water policy is often described within a narrative of progress in which there is a linear trajectory towards the current 'mature arrangements'. But the dam builders have not ensured a supply of water for agricultural production. Their water projects have failed. The result is what Bruno Latour describes as proliferation: the further entanglement of the fortunes of already connected people and ecologies. In this proliferation, government agencies and volunteer groups are now kept busy devising more and more elaborate ways to address the devastation: removing redundant weirs, restocking native fish, building fish passages, conserving waterbird habitats, protecting wetlands, restoring riparian vegetation, and resnagging the rivers with trees. These are attempts to replicate the work that used to be done by the rivers. Yet, this 'ecological support work' is just tinkering with a system that continues to be based on enormous extractions of water for irrigation. Governments will not be able to address what is important while they still back the powerful tools of modernity and deny their impacts. Because water is such a powerfully connected life force, this failure has profound consequences for all who rely on the life of the inland river country.

CHAPTER 3

Connectivity, loss and resilience

A thing is right when it tends to preserve the integrity, stability, and beauty of the biotic community. It is wrong when it tends otherwise.

Aldo Leopold, forest ecologist (1949)

The traditional owners from the inland river country express ecological destruction in terms of their very being. Their concerns are more than a warning system about changes to river health; the traditional owners offer a deep understanding of ethical relationships with the rivers, the ecological life along the rivers, and the river country. Often referring to it as 'connectedness' or 'being connected', the traditional owners appreciate the rivers as being embedded within networks of relationships. This is in stark contrast with modern water management that interprets river water as a mute resource — undifferentiated and unconnected. For the traditional owners, nature is not just nature, it is also culture; and culture is not just culture, it is also nature. Following their lead, I argue for an expanded concept of ecology as an answer to the limitations of the binary focus of the moderns, described here as 'connectivity'.

IMAGINING AN EXPANDED ECOLOGY

In academic literature the concept of connectivity has its roots in the analytical work conducted by ecologists, and focuses on relationships between all species and their environment. Ecologists use connectivity to describe the way in which animals and plants live in interconnected relationships across multiple spatial and temporal scales. For ecologists, connectivity is 'the ease with which organisms, matter or energy traverse the ecotones between adjacent ecological units' (Ward et al. 1999, p. 129). The importance of the connection is emphasised rather than the substance of that which is connected.

In landscape connectivity, spatial structures and habitat patches provide different species with different opportunities for movement. In floodplain river ecosystems, advancing and receding waters create a shifting mosaic of habitat patches. Hydrological connectivity maintains a diversity of connected ecological zones over time and space. This analytical work has critically countered narrow perceptions of water as an abstract resource for consumption, to focus on freshwater as a life force. However, this ecological connectively is often represented as external, where humans are the only animals not included, much like 'wilderness' thinking. An expanded connectivity employs the holistic thinking of what Bruno Latour has called amodern knowledges (see Chapter 1), including Indigenous peoples' knowledge, to (re)position humans *within* a web of life-sustaining relationships.

The conceptual habit of analytically removing humans from their environment is part of the nature/culture dualism. The problem with making distinctions such as nature/culture is not that distinctions are identified, but how the distinctions are organised into binaries. Further, modern philosophers engaging with Cartesian dualism have hyper-extended the binaries into oppositional relationships. If humans are rational, then nature is mindless; if humans are active, then nature is passive (Rose 2007b). The distinction is transformed into an insurmountable tension that cannot be resolved (Latour 2001, p. 58).

Environmental philosopher Val Plumwood described this modern habit as 'hyper-separation', and Bruno Latour describes it as 'hyper-incommensurability' (Plumwood 2002, p. 49; Latour 2001, p. 61). The result for the moderns is a highly flawed perspective, which both increases human power to transform nature, and limits human capacity to respond to ecological devastation. As Val argued (Plumwood 2002, p. 9):

> [When] we hyper-separate ourselves from nature and reduce it conceptually in order to justify domination, we not only lose the ability to empathise and to see the non-human sphere in ethical terms, but also get a false sense of our own character and location that includes an illusory sense of autonomy.

This is also a structuring of hierarchical power relations, with humans assumed to be dominating nature.

Central to this power to transform nature is the assumption in modern thinking that modern knowledge is universal and thus can be applied anywhere. The moderns recruit scientific knowledge to this work. Their 'universal knowledge' not only makes authoritative representations of the real world, but replaces the real world with the representations. Likewise, Donna Haraway has argued that representations of 'the living world itself' in the discourse of biology have come to substitute what really exists for that representation (Haraway 1992, p. 298).

We should focus on the use of scientific knowledge by governments to understand the dramatic transformations of nature in recent times. Scientific knowledge is a lot more complex than the interpretations given to it by governments seeking to 'solve' environmental management. In their use of universal science, governments have not valued the scientific work that reveals the complexity of ecological life (Rose 2007a). Governments tend to maintain the modern illusions. In response, many academics from the sciences and the humanities have counter-theorised relationships and connections, such as complexity theorists, actor-network theorists and Deleuzians (after philosopher Gilles Deleuze; see for example, Bonta 2005; Bonta & Protevi 2004; Murdoch 1997). This is a period of dynamic theoretical change in response to ecological devastation.

Instead of getting caught up with representations that authorise modern knowledge as universals, Donna Haraway argues for knowledge that comes out of particular places. This she calls 'situated knowledges'. For Donna, accounts of the 'real' world are not dependent on universally held laws but on conversations held between actors of many different forms (Haraway 1988, pp. 581, 593). This is not a retreat to postmodern relativism, within which only the local can be known by a local. Scientific knowledge is not to be abandoned. Rather, as anthropologist Ben Smith argues, we need to recognise the local assumptions in scientific knowledge and the 'conflation of instrumental technique with the "real" it describes' (Smith 2007, p. 82).

The expanded connectivity I describe embraces the world as full of life and agency beyond human activity. We need a particular kind of 'presence-in-the-world' to enable us to encounter this connectivity, as Debbie Rose has written (Rose 2004b, pp. 213–14):

> I mean this presence to be situated in history and in place, and I mean it to be available to social and ecological 'others'. Attentive and alert to the here and now of life, the kind of presence I argue for, in addition to being situated and available, is relational, connective, mutual, and committed.

We need to move beyond considerations of a separate, subordinate nature, to consider living ethical engagements within a dynamic nature.

Crucially, ecologists are theorising about connectivity as a subjective experience for all species. Connectivity for one species will be different from another, as each species has a unique experience of living in and perceiving their ecological niche or 'umwelt' (Allen & Hoekstra 1992, p. 169; Hoffmeyer 1997, p. 54; Manning et al. 2004, p. 622–3.) Thus, there are as many understandings of the world as there are species. Rather than positing these subjectivities as a barrier to action and understanding, umwelt reinforces the amodern perspective that there can be no single understanding of nature. This scientific work on subjectivity and agency brings sentience to these newly acknowledged actors. Sentience is the capacity to have feelings and/or feel sensations. In scientific circles, work engaged with sentient beings was pioneered in the 1960s by primatologists, including Jane Goodall and her collaboration with chimpanzees in Africa. A specific branch of science now focuses on relationships between people and animals: anthrozoology.

Anthrozoologists have made persuasive arguments that animals have feelings and thus are sentient. These arguments are extended by ethnographer David Anderson who has theorised the myriad of solidarities and obligations between people and places and animals as a 'sentient ecology' (Anderson 2000, p. 116). This ecology brings us into communicative relationships with the ecological world, and extends the concept of personhood to all ecological life, not just animals. Anthropologist Timothy Ingold discusses how these communicative relationships evoke feelings of care, love and attachment towards the environment, similar to those one feels towards another person (Ingold 2000, pp. 69, 76). The concept of a sentient ecology is important because it can help us reconceptualise our relationships with ecological life away from causal relationships of power. A sentient ecology establishes an emotional and ethical context for our ecological relationships.

Understanding a sentient ecology is critical to reorienting our engagement with water issues. According to anthropologist Gregory Bateson, our survival depends on our understanding that we are coupled to our conceptualisation of our ecological relationships, and our ways of thinking and acting on them (Harries-Jones 1995, p. 8). Humans lose water constantly — we are not water tight — and we need to replenish ourselves or die. This is a 'hydro-contract', an inescapable biospheric life support which we need to work with to maintain 'hydro-harmonies' (Warshall 2002, pp. 42–3). The conceptualisation of water as an abstract unconnected resource denies these connections between water and human survival.

To deepen our understanding of an expanded connectivity, we can explore conceptual frameworks that are not entrenched in a history of dualism. The traditional owners from the Murray River have inherited Aboriginal knowledge frameworks that have only recently been transforming through engagements with dualistic knowledge traditions. The traditional owners speak of a connectivity that goes beyond food-web dependencies to include stories, histories, feelings, shared responsibilities and respect.

WATER CONNECTIONS: COMPLEXITY AND THE ECOLOGY OF LIFE

The Murray River Elders reminisce about times past when their lives were connected to the rivers through the essential act of drinking the water and eating the plants and animals that also lived by the river. In South Australia, Ngarrindjeri Elders Matt Rigney, Richard Hunter and Agnes Rigney spoke to me about how they were able

3. CONNECTIVITY, LOSS AND RESILIENCE

to drink water straight from the Murray River and its lakes when they were young. In those days, said Richard and Agnes, the water was so clear you could see the bottom of the river. Agnes described to me how she grew up in a 'semi-traditional' lifestyle next to the Murray River at the Swan Reach mission. Here, the Murray supplemented mission food with fresh water, fish, yabbies and waterbirds. This subsistence economy was mixed with the welfare economy provided by the missions and the traditional owners' involvement with the market economy. Upstream in the Barmah Forest, in New South Wales and Victoria, Yorta Yorta Elder Henry Atkinson told me how his mother's father and his father were able to live off and make a living from fishing for native fish, mussels, Murray crayfish and turtles.

As Agnes Rigney talked about her experiences she expanded connectivity to include a merging of the river with her own body. Agnes spoke about this as she explained why she continues to live near to the Murray today:

> I don't think I can be far away from the river because the river I believe it is in my blood. It is a part of me. I was born on the river. I have lived on the river all of my life and I am an Elder now. I wouldn't be happy too far away from the river … We are all part of the food chain, and that's why I say I feel a part of it — well I am … The river gave us life, the river fed us. [21 July 2004]

Agnes is expressing connectivity as an embodied experience. This relationship includes her daily lived experience next to the rivers, full of sensory encounters such as sight, smell and touch, as part of her intimacy with the river. Such comments can also be heard from settler Australians who have long and intense relationships with an environment. They talk about the way the place gets into their bones, or into their blood (Rose 2004a).

Agnes places herself *within* the relationship of connectivity created by the Murray River. This is a perspective that moves beyond an understanding of the world as separated into spheres of human and natural, to an understanding of a world in which our being and the environment are bound together. This is not taking a view of the world, but is 'taking a view *in* it' (Ingold 2000, p. 42. original emphasis). This relationship with the environment is a dynamic experience of life and survival. It is not a simple addition of our being to the environment, rather it is an acknowledgement that we affect the environment just as the environment affects us. We do not simply live together, side by side, as a matter of coincidence, but our form and being are interconnected. Our being and the environment are active, alive, and respond to each other through multiple fields of relation, and these interactions influence the form of the relations (ibid., p. 19).

This experience is not exclusively human; it is an experience shared by all life forms. Different species have developed close symbiotic relationships in which they are co-dependent for survival. For example, lichen is not a single organism but a combination of fungus and algae growing together in symbiosis — a mutually beneficial connectivity. Other connectivities may be beneficial in primarily one direction, as when insects lay their eggs on eucalyptus leaves. Such seemingly small connectivities can be expressed on much larger scales, as their lives encounter innumerable links with other lives, including the exchange of energy in food-webs. The environment and living things make exchanges through webs, clusters, knots, loops, ripples, waves and curves. Gaps in these interactions permit the creation of distinctions and differences (Harries-Jones 1995, p. 14). For example, the furthest reach of the floodwaters creates a distinct ecological boundary. Acknowledgement of these holistic and diversely interconnected relationships is an approach to ecology that Timothy Ingold has summarised as 'the-whole-organism-in-its environment'; he calls this 'the ecology of life' (Ingold 2000, pp. 18–19).

Connectivity not only ensures that life-benefits ramify, but can also become conduits for damage. This has happened with the 'white death' that is salinisation, which occurs through the rupturing of certain connectivities while others remain. Salinisation is a process wherein agricultural land becomes so salty that it cannot support life. The salt occurs naturally in the landscape and it is dissolved and brought to the surface when the watertable rises (Proust 2003, p. 39). The watertable is rising on farm lands because of excess irrigation water being added to existing groundwater and/or the clearing of deep-rooted vegetation which previously regulated the watertable by drawing down the groundwater. When the water that rises from below evaporates from the surface, the salt that is carried with it is left behind.

Agnes brings us into this world of salt and water by talking about the river flowing through her veins. Agnes's world view reveals to us what Gregory Bateson meant when he said that an organism that destroys its own environment is committing suicide (cited in Rose 2001, p. 3). Agnes places her life within her relationship with the river.

SENTIENT BEINGS, SENTIENT ECOLOGY

When Agnes says, 'the river gave us life, the river fed us', she is describing that relationship as a caring and giving one. It is this appreciation of the river as a sentient being that brings agency into focus. Indeed, many of the traditional owners attribute the capacity of the river to sustain life to a life force that is the river itself. This capacity for feelings can be extended to all things, and in the broadest sense country is appreciated as being alive and having the capacity to act. Aboriginal understandings perceive country as a 'nourishing terrain' that both gives and receives life, and is lived in and lived with (Rose 1996, p. 7).

Being responsive to extra-human agency is something Yorta Yorta man Lee Joachim talked to me about at length. Lee understands his relationship with the Murray River as a relationship held between sentient beings:

> The importance of the river is to ensure that it is seen as a continuing living being. That it is respected like any other person should be respected. It has got the ability to cleanse itself. It has got the ability to nurture itself. And it has got the ability to ensure that the life that it touches upon also has an ongoing process. [25 June 2004]

Lee brings the river into the foreground. This perspective moves away from a world where humans transcend nature. By extending the recognition of power and agency to all living things, Lee grasps a dynamic world in which humans and other beings participate.

The importance of extra-human actors in the world was something I was introduced to by the traditional owners. Early on in my research for this book, Yorta Yorta woman Monica Morgan critiqued the United Nations' (UN) approach to protect a human right to water by saying that such approaches went 'too much the other way'(comment in a MLDRIN meeting at Swan Hill in September 2003). She explained that the UN approach misses the point about respecting country by recognising the importance of water bodies only in terms of human needs (UNCESCR 2002). Monica critiqued this modern analysis that denies the agency of living beings other than humans. To extend her idea: if nature is just matter then nature is neither hostile nor friendly but rather indifferent to our interests; it is thus possible to exploit nature without regard for its agency or interests (Mathews 1994, pp. 14, 32). Such perspectives enable moderns to transform nature without considering the ethical consequences.

Lee linked his argument that the river has its own agency to his other argument that this agency demands its own respect. By appreciating the river as a sentient being, Lee wishes to inspire in other people feelings of empathy, care and respect towards the river. Lee has argued that this understanding is critical to transforming the way the river is appreciated in Australia:

> They must see that there is a connection to everyone's life through the rivers and through the environment attached, that is an ongoing care and recycling of themselves and a continuation of life within that. But they just don't seem to be open to the fact that the river can speak for itself, and the country can speak for itself. [25 June 2004]

This is a communicative relationship, and Lee brings all people, Aboriginal and settlers, into this relationship. Our survival depends on our ability to respond to this communication. I asked Lee what he meant by the river and the country being able to speak, and Lee replied that the crickets and the frogs did not make as much noise anymore, 'those noises that tell you that they are alive and well', and the animals and the fish had disappeared. Their absence is now communicated by silence. Lee is describing a sentient ecology where listening is part of being connected. These communicative relationships are critical to informing our responses to ecological devastation. Because, as Debbie Rose has said, 'to hear is to witness; to witness is to become entangled' (Rose 2004b, p. 213). For Lee, country is speaking loudly to him about connectivity and loss, and Lee is ethically compelled to respond.

WHEN NATURE/CULTURE MEETS ANCESTRAL CREATORS

For the traditional owners such as Lee, country is not inert, non-responsive matter; rather, country is communicative and has agency. Country is an active participant in the contemporary water debates in the Murray–Darling Basin. This is a source of authority that the moderns had not conceived.

One of the most significant aspects of connectivity is that it offers an alternative to modern knowledge frameworks that hyper-separate nature and culture. Similar to Bruno Latour's arguments, anthropologist Marilyn Strathern described how the nature/culture distinction is part of the moderns' habit of engaging with concepts as oppositional relationships (Strathern 1980, pp. 186, 179). She points to the lack of consistency in the images of nature and culture in western constructs, as evidence of how the distinction is struggled with in western thought (Strathern 1980, p. 190), and writes (ibid., p. 177):

> There is no such thing as nature or culture. Each is a highly relativised concept whose ultimate signification must be derived from its place within a specific metaphysics. No single meaning can in fact be given to nature or culture in western thought; there is no consistent dichotomy, only a matrix of contrasts.

The separation of nature and culture is reflected in Commonwealth Government heritage legislation which is designed to protect our natural *or* cultural heritage (*Environmental Protection and Biodiversity Conservation Act 1999* (Cwlth), s.12(3)). At a workshop I attended for heritage practitioners and academics in Canberra, one participant was frustrated by the complexity of the nature/culture categorical division and implored a presenter 'Just tell us how to better separate them' (Anon. 2005a). The nature/culture separation is persistent, and is infused through all sorts of other legislation. For example, as part of their incorporation under the *Corporations Act 2001* (Cwlth), the alliance of traditional owners the Murray Lower Darling Rivers Indigenous Nations (MLDRIN) set up separate environmental and cultural trusts because the legislation makes it extremely difficult to set up a joint fund.

Seeing the world as connectivity addresses the separation of culture and nature because connectivity places people (and their culture) *within* relationships with the environment (or natural world). Our minds are not suspended from a reality that we then must try to understand externally, rather, our whole selves experience and apprehend the world as a part of living in it. This is not to say that we cannot objectify the world. As Paul Sillitoe has argued (Sillitoe 2007, p. 3):

> All humans are capable of abstract thought and have notions of causality, that they can suspend prior beliefs and will revise these if evidence suggests that they are wrong, even if counter-intuitive … All cultures accumulate and interpret knowledge rationally according to their value codes, although until we appreciate these latter it may seem otherwise.

What connectivity-thinking additionally acknowledges are our sensory experiences. Connectivity breaks from abstract thought to acknowledge the palpable links that exist between mind and body. The intellect is inseparable to our feelings and experiences of living in a sensory world (Harries-Jones 1995, p. 5).

The postmoderns challenge such embodied connections because, to reiterate their argument, we can only engage with a world of images. Similarly, some anthropologists argue that the world is socially constructed by humans, and as humans we can only know our perceptions and can never know the world as it is. As anthropologist Roland Littlewood has argued (Littlewood 1996, pp. 122–3):

> We determine only in part what we call our environment, but we determine our experience of it, our human world. 'It' determines us, we 'are' it, but this 'it' is only an 'it' through human procedures, shared with our fellows.

Not all anthropologists agree with Roland and the nature/culture divide, and this is actively debated in anthropology (Ingold 1996a). Timothy Ingold has counter-argued that the world is not 'free-floating' and separate to us, but is 'coming into being through the activities of *all* living agencies' (Ingold 1996b, pp.139, 141; original emphasis).

When Matt Rigney looks out in Ngarrindjeri country at the Murray River he can see where his ancestral creator Ngurunderi chased the giant *pondee* down the small stream that was the Murray River back then. Matt connects this creation story to the ecological life of country, his position within this life, and to his rights and responsibilities to look after country:

> [This story] tells us how our country was created and what was the purpose of the creation — it was to sustain life, to give life, and to create an environment that sustains us in that way. Like the bird life, the animal life, the plant life. So I have an inherent right, a cultural right, and a responsibility to make sure that those things are maintained and continue to survive and live for the duration of time. It makes me who I am. Ngarrindjeri man. It is my responsibility as a Ngarrindjeri man to make sure these things happen, and that our culture and spirituality is not disconnected from the river and the waterways. [24 July 2004]

In the landscape of the Murray River, the ancestral past, today's world, and Matt's identity all meet and are connected. For Matt this is not just knowledge held in his mind, but knowledge held in country and experienced through the senses. Timothy Ingold describes this perspective as the long-held sensitivities and orientations of living within an environment (Ingold 2000, p. 25).

Alternatively, this Ngarrindjeri creation story could be described as a cultural perspective that comes from a culturally constructed discursive world and is projected onto the natural world. That is, the creation story has its origins in discussions held in Ngarrindjeri cultural life and is a Ngarrindjeri-constructed interpretation of the natural world. In this view, the connections that Matt observes are romantic or religious, but they are not 'real' or scientific.

This perspective relies on the separation of culture from nature, of mind from body, of spirit from matter, and of subject from object. Timothy Ingold has argued that these separations are impossible to achieve. In order to place this cultural perspective onto the real or objective world, Matt would have to be able to step outside of the environment within which he lives. For an observer to then view that this creation story is a cultural perspective and not part of the environment, the observer would have to take a further step outside of both Matt and the environment (Ingold 2000, p. 14). Matt, however, is not taking any such steps and instead is embedding himself in relationships with the natural world and identifying the causal and connective relationships:

> We are of these waters, and the River Murray and the Darling and all of its estuaries are the veins within our body. You want to plug one up, we become sick. And we are getting sick as human beings because our waterways are not clean. So it is not sustaining us as it was meant to by the creators of our world. [24 July 2004]

Dreaming stories are the epitome of a sentient ecology. They describe country not as inert resources but as a lively narrative with immanent ancestral beings. This is a way of understanding country as part of cosmic energy, within which the traditional owners have certain responsibilities. This connectivity brings into the foreground the life-sustaining co-dependences held between all living agencies, and the traditional owners look to country to tell them about their own lives. A healthy country tells the traditional owners that the ancestors are happy, and that they are managing their rights and responsibilities. But if country is sick, then so are the traditional owners — sick with the diminishment of life.

ECOLOGICAL ENTANGLEMENTS, CULTURAL LIVING AND DISPOSSESSION

In this understanding of ecology as sentient and experiential, changes to river management over the last 60 years have generated traumatic repercussions for the traditional owners of the river country. As noted earlier, with the extensive settlement and agricultural industry in their traditional country, continuing access to the river has been integral to maintaining connections to country and their traditional identity. The impacts of ecological destruction have now caused a new wave of dispossession (Weir 2007b). The changes are noted in the Elders' dismay at the scale and speed of the destruction; as Mutti Mutti Elder Mary Pappin said (Pappin 2004):

> Such a short space of time! I can't take my grandchildren down to my favourite fishing spots and do what I used to do.

With ecological destruction, the traditional owners have lost opportunities to connect with and reaffirm relationships with country and with each other. This is a heavy loss because of the unique specificity in the relationships traditional owners hold with their country.

Monica Morgan spoke to me about the river and the river life as experienced in her childhood in the 1960s in Yorta Yorta country. When Monica was a child her Elders taught her the ecological cues that would tell her when swan eggs would be available up in the Barmah–Millewa lakes:

> There was life … You'd sit there and they'd say, 'Oh well the duckweed is coming down; that means the swan eggs are ready to go and be collected up in the lakes.' So there were seasons happening. [1 July 2004]

The Yorta Yorta Elders passed their knowledge on to Monica, and now she can pass that knowledge on to her own children. However, this is a memory of the past. Today the flow of the river has been dramatically modified, the Murray now floods out of season, and both the swans and the duckweed are rarely sighted. This is a real concern for Monica:

> So if I am seeing in just a short time twenty, thirty years, the disappearance of things that I took for granted, and were a real reflection of nature, and also governed my life cycles, then what is going to be left for our children? [1 July 2004]

Agnes Rigney directly links this loss of life by the river to the loss of what she calls 'cultural living'. My understanding, from listening to Agnes, is that cultural living reaffirms continuities with the ecological world through the practising and passing on of cultural knowledge and experience. For Agnes, this cultural living is connected to the river:

> I remember as a kid growing up in Loxton how clear the river was, the water was, and my father actually making us spears from bamboo. And we used to walk down to the river and we used to spear the fish. And it is just sad what's happened to it now. That was a part of cultural living, connected to the river, that we can't really practise anymore. [21 July 2004]

Living in, surviving on, experiencing and enjoying country is difficult when the life of that country is stressed, sick or absent. These connectivities are not just held between the traditional owners and their country, but between the traditional owners and their ancestral beings, ancestral creators, old people, and their children, grandchildren and future generations.

Since the expansion of modern water management in the mid-twentieth century, bush foods and medicines have become harder and harder to find. As cultural living is a practice rather than a theory, cultural living ends where it is difficult or impossible to perform these practices. The rupturing of this living practice breaks the connectivities that the traditional owners are responsible for, thus breaking cultural living.

Ethnoecologist M Kat Anderson has described the adverse impacts of the loss of ecosystems on the wellbeing of Indigenous peoples in North America, and summarises 'at least' five categories of potential loss: ethnic identity, economic independence, health, religious freedoms and Indigenous knowledge about ecology (Anderson 1997, pp. 16–20). Kat Anderson describes how the local extinction of plants and animals has ramifications in terms of identity because the unique ecology of a place is woven into the identity of the people who hold long-term associations with that place. In turn, the ecological resources of a place also sustained the economy of the local Indigenous people, and with the destruction or mining of these resources those local economies are devastated. This also has health implications, as the degradation of traditional lands and waters increases exposure to toxins and disease. Furthermore, the loss of species and the transformation of landscapes affect the opportunity for Indigenous people to exercise their religion because religious ceremonies are intimately connected to local species and places. Finally, Kat Anderson points out that the perpetuation of Indigenous knowledge about ecology requires access to that ecology. As ecologist Robert Pyle has described it, the local loss of species is 'the extinction of experience' (Pyle 1992). When a species becomes extinct, all people lose direct, personal contact with that species. As time passes, the next generation does not have the opportunity to know the species, to interact with and identify with it. In the Murray–Darling Basin, this concept can be extended beyond species to wetlands, rivers and landscapes.

When I asked Mutti Mutti Elder Mary Pappin what issues she thought were most important for her country, she spoke about protecting Aboriginal sites for teaching the next generation about their culture:

> because if a site is destroyed, they can't identify with that site. Of course, if you can't identify or teach from a site you might as well go back to look for it in a book! [22 July 2004]

In this scenario the role of interpreting Mutti Mutti culture is shifted to a book, which then becomes the authority. In the book the knowledge is held in a static form and is coded differently by being abstracted from the specific relationships held in place. Knowledge is no longer transmitted as an embodied, situated experience, but is separated, producing the mind/matter binary.

The over-extraction of water from the inland rivers is so great that it is being experienced by the traditional owners as a contemporary dispossession from their country. Country is not the alive, dynamic place the Elders grew up with. This loss is a wave of dispossession that goes to the very heart of the traditional owners, following the webs, clusters and loops that entangle the traditional owners within sentient ecologies. It reaches into the very source of authority and knowledge that the Elders want to keep alive for the next generation. Ethical relationships of care are broken down. Only by understanding and valuing this as situated knowledge that is embedded in the world is it possible to understand the reverberations of this loss.

THE DEATH OF CONNECTIVITY

The change in the rivers over the lifetime of the Elders has been remarkable. They were once able to drink water straight from the rivers, and now even the Murray crayfish struggles to survive. It is no longer possible to collect duck eggs with an awareness of the changing seasons. It is no longer possible to see the river life thriving and to acknowledge that life as central to the sustenance of one's own life.

The ecological loss that has occurred is significant with respect to the range of life that has been destroyed, the short timeframe, and the wide-ranging scale of this destruction. The degradation of country is expressed by Ngarrindjeri Elder Agnes Rigney (Lower Murray in South Australia), Ngarrindjeri Elder Matt Rigney (whose country is at the Murray Mouth and Coorong), and by Yorta Yorta Elder Henry Atkinson (whose country is that part of the Murray centred on the Barmah–Millewa wetlands east of Deniliquin).

> Agnes Rigney:
>
> It is not alive today; it is a dead river. Not only from just looking at it, but what it produces. Yes, I've seen the changes. I've seen the time when the river did produce for us well, when the river was clean, you could see the bottom of it. But to see it now it makes you wonder how anything could live in it actually, like the fish and the micro-organisms and all that. [21 July 2004]
>
> Matt Rigney:
>
> Now I look at it and I see a rubbish tip. Just surviving, that's it. You can't drink it. You can't swim in it. I'm talking about Lake Alexandria of course and Lake Albert, too scared to go for a swim in summer because of blue-green algae. [24 July 2004]
>
> Henry Atkinson:
>
> The Murray River is being treated as a drain. The quality of the water is what you would find in a gutter. In fact, I would not go swimming in the river in Echuca as I used to anymore, there's no way. [7 August 2004]

For these Elders — for whom the river is so important, is described as part of their bodies, their blood, their veins, part of their survival as a people, part of their cultural heritage, their ancestral inheritance, economy and cultural living — to describe the river as a rubbish tip, as a drain, a gutter, and as a dead river, signifies the intensity of the feelings about this loss and the desperation of the situation. It is a death that reaches through time to dissociate them from their ancestral past and to devastate the future.

3. CONNECTIVITY, LOSS AND RESILIENCE

Death and dying is a part of the recycling of life by the rivers, but this is a death that breaks down the life-supporting connectivities. In this death there is no connection to the regeneration of life. There is no joy to be had from places without life.

Lee Joachim put it this way when he talked about what has happened to the Murray and how it relates to the Yorta Yorta people:

> We just don't all come from Adam and Eve. We come from the simple dirt that we walk upon. And our spirits, and our Baiame, our makers, it's all interconnected there. And people don't even show respect, you know, for that. If our river and environment is dying, then I believe that we as a people are also dying. [25 June 2004]

For Lee, the traditional owners and their country are bound together so intimately that they will survive together and they will die together. Lee believes that if the river dies, then the Yorta Yorta as a people are also dying. I interpret this as meaning that the people who make up the Yorta Yorta will continue to live, but no longer with any connection to their traditional country, and thus no longer truly Yorta Yorta. Traditional ownership thus becomes an abstraction rather than a lived reality. Country also becomes an abstraction; a concept rather than a sensory experience.

Henry Atkinson is so incensed by what has happened to Yorta Yorta country that he suggested to me that people 'can pack up and go and do whatever they like, and leave the river and land a mess'. If the abuse continues, Henry has argued, eventually all the water in the Murray–Darling Basin will dry up and all the land will become a salt basin. Henry is incensed because he feels that people do not care for country, and do not acknowledge that the end point of treating water only as a consumptive resource is the annihilation of country. The narrative of harnessed water for agricultural productivity is destroying the narratives of his people and his country, along with the destruction of the life of the rivers and the land.

The depth of this grief for the loss of country, often expressed to me in anger, is understood within a sentient ecology, where the river is recognised as an actor in sustaining the life of country, a central part of the traditional owner's life experiences and identity, and part of a communicative relationship with the life of country. Historian Peter Read coins the term 'place-bereavement' to describe grief for the loss of a place experienced by both Indigenous and non-Indigenous people, which he says is akin to the grief for a loved person (Read 1996, p. 19). However, the situation that the traditional owners are describing is more than the loss of one place, indeed it is more than the loss of the full expanse of their traditional country. This is about the loss of beneficial connectivities along the entire length of the rivers. This is also the loss of relationships held throughout the inland river country, which are at the base of the formation of MLDRIN. This loss is not about the pragmatic choice of foregoing one place, to support the continuation of another. This is about a loss of connectivity across a large part of the Australian continent.

REGRET, DREAMS, LOVE AND REFLECTION

The power and agency of the traditional owners are a necessary part of an expanded understanding of connectivity. Passion, anger, disappointment, regret, beauty, love, care, memory, dream, reflection — these are all sentient engagements with country. Country evokes feelings, and the traditional owners value these emotional connections. Beautiful healthy living landscapes engender positive responses. Places that are destroyed are reacted to with anger or regret. The feelings the traditional owners describe in response to ecological destruction cannot be disengaged from that destruction.

Wiradjuri Elder Tony Peachey spoke to me with regret about his involvement in the clearing of vegetation on a property in Wiradjuri country when he was much younger. Peachey took the job out of economic necessity to support his family. Thinking back on it, Peachey expressed how depressed it makes him feel:

> That's not a real good thing to think about. When you go back there now every property, all around there, has done virtually the same thing. When I was a kid I used to go chasing pigs through there on horseback with my cousins. If you go out there now and just stand up on the big reservoir and you look around for miles and you can't even see a tree. It's just bare. [29 October 2004]

At the time it was a job, but now, standing on top of the big reservoir, Peachey can see that his labour was part of the far-reaching transformation of his country into a bare place. While cutting down trees in itself is not inconsistent with philosophical traditions of connectivity, cutting down *all* the trees completely contradicts a philosophy that respects the living world. Peachey realises that his contribution was a small part of a greater framework of capitalist production that seeks to intentionally dominate and transform the world for market return. It is the excessively expansionist model of economic growth that is grasped as problematic, rather than economic needs and desires in themselves.

Wiradjuri Elder Ramsay Freeman spoke with me about the nostalgia he feels for times past for certain places in his country. Near to Ramsay's home in Tumut is the Blowering Dam, a reservoir for the Snowy Mountains Hydro-Electric Scheme. In the 1960s Ramsay worked for Theiss Brothers constructing the road up to the dam. When MLDRIN held a meeting in Tumut in December 2003, Ramsay took

3. CONNECTIVITY, LOSS AND RESILIENCE

The usually drowned Yellowin Bay emerges when the water level drops at the Blowering Dam, Tumut, New South Wales. (See also colour plate, between pp. 80–1.)

everyone up into the mountains and pointed out Yellowin Bay, at the northern end of the Blowering Dam. Here, Aboriginal people used to meet for the Bogong moth feasts (Flood 1976). At the time of our interview, the drought had dropped water-storage levels at Blowering to 8 per cent of capacity (Anon 2005b), re-revealing Yellowin Bay and other drowned places (see figure above). Numerous stone tools, flakes and cores were washed up on the edge of the shore, showing the activity of Aboriginal people in earlier times. When he can, Ramsay likes to go there and sit and think about when everyone met up to get the Bogong moths, and how the way up into the mountains, the saddle, is still there. Drought reveals the old places in the Blowering and brings important sites back into view, making memories visible and uncovering the violence of the current imposed 'order'. (Read (1996) has also discussed dams and loss.)

Further downstream in South Australia, Ngarrindjeri Elder Richard Hunter told me that he could see beyond the loss that is the damaged landscape to the beauty that is still there. Richard's home is perched on the top of the gorge made by the Murray River, with broad views of the dramatic orange cliffs that follow the river's bend to the small town of Nildottie. Richard has lived in this area all his life, growing up on the mission further upstream at Swan Reach. Richard told me of his source of strength:

> Well, I have got a good dream of what they done back there. But I know I'll never get to show it to you today out there. But I still sit there with the beauty that I know is there, and I'm never going to lose that … That is the beauty of the land. We can see the beauty, what the rivers were before, and the tall trees. They talk about the damage that has happened now, but we were still taught how to have the skill to see the beauty that is there. Even though it is not there physically, but it is, because you know what it was like before. Before Europeans come here. [23 July 2003]

My understanding is that Richard is telling us that some may think that we are heading towards the death of connectivity, but that connectivity of life is still there. This life has an enduring presence that is more than physicality, it is a presence that cannot be killed or destroyed. Richard can go there in his mind's eye, seeing through today's salty silted Murray River to the beauty of a living, enduring river. The power of the river is a force of life beyond physical manifestations. Perhaps this is a reference to the Dreaming, although Richard did not say so to me. The Dreaming is the time of creation, but this sacred time is not situated in the past, it is ongoing, it is 'everywhen' (Stanner 1989, p. 228). Or perhaps this is a vision of what can be — neither the past nor future — a potential.

This is the vision of one man about the resilience of the river country. Such visions sustain people in the work of restoring the river country. MLDRIN is part of this remix: a belief beyond the death of connectivity to the resilience of the river country.

RESILIENCE

Resilience is a concept described by ecologists as the ability of ecosystems to be self-ordering and self-repairing, as each living thing has its own will to flourish. For humans, resilience has a similar meaning. It is also about the capacity to cope with catastrophe, to find ways to continue life-sustaining relationships within their environments (Rose 2004b, pp. 6–7). In their work to respond to ecological crisis the traditional owners reveal their resilience, interconnected with their belief in the resilience of the river country. They are open to the importance of these relationships and the power of the river as the key agency in river restoration.

3. CONNECTIVITY, LOSS AND RESILIENCE

Down by the river, the traditional owners continue to find inspiration in the flow of the water. Many of the traditional owners spoke of the calming and soothing effect of being near water, even though the rivers are suffering. For Peachey, being out on the river in a boat is the place where he gets his best ideas, and his sister makes him take a notebook so he can record his ideas and not leave them out on the water. Mary Pappin can imagine the Dreamtime when sitting by the river. The water, river, gumtrees and wildlife make her feel that she is 'sitting right in it', as she says, 'without the interruption of white people'. The loss that has occurred along the rivers is lamented but is not used as an excuse to succumb to despair. Rather, the river is a motivating force and the traditional owners draw succour from country. Positive reports of river health are noted as people look to country to show its ability to sustain itself. Up north in Wiradjuri country, Peachey reports that the magpie geese have returned to the Macquarie Marshes. Down at the Coorong, Matt Rigney expresses his joy about the return of a musk duck.

Some of the traditional owners spoke to me about a desire for a great flood, a flush of water going through the entire river system. This flood would wipe out the weeds, improve the water quality, and give the river the chance to regulate its own health. The longed-for flood would be an expression of the freedom, energy and life-giving power of the river. The flood would return us to the times of droughts and floods, when the river was free of dams and weirs. The flood's watery blanket would cover all places and seep in, returning country to the rivers' domain.

If the advocates for sustainability within the government are looking for ways to foster long-term values in the broader community, to gain public commitment to the work of repairing the river country, then understanding the relationships, ethics, and insights of the traditional owners is a valuable place to start. These are different ways of learning and communicating knowledge that form loops of connection rather than single objective trajectories.

CHAPTER 4

Setting the negotiation table

Our feet all rest on the same earth.[1]
Donald Grinde, Yamasee Indian and historian, and
Bruce Johansen, communications academic

The traditional owners bring the language of connectivity to their negotiations with governments, but these governments have a long history of governing with the language and concepts of modern knowledge. Governments whose governing principles are founded in universalising modern knowledge fail to recognise the cultural biases within their laws, policies and programs. Because of this, so-called 'participatory' approaches to include Indigenous people in modern laws, policies and programs do not engage with amodern perspectives. When invited to participate, Indigenous people find themselves faced with the 'choice' of either co-option or marginalisation (Howitt & Suchet-Pearson 2006; Tully 2004a, pp. 57–8). Indigenous people who oppose these prejudicial processes can then be characterised by governments as unreasonable radicals, or simply cut out of the participation process altogether.

Modern approaches challenge Indigenous people by determining what is possible and/or appropriate for an Indigenous person to negotiate, though Indigenous people also work with modern knowledge to make their arguments. Indigenous people have adapted and translated modern knowledge as part of their intercultural experience, and modern philosophies have consequences for the way Indigenous people perceive themselves. In what follows, I shall show how the MLDRIN delegates theorise themselves as 'traditional' within contemporary Australia.

Modern thinking can be oppressive for Indigenous people, but it also has the power to liberate. Instead of rejecting modern thinking, we need to re-situate it within broader knowledge frameworks. To do this, as Bruno Latour has argued, we have to acknowledge the work of the moderns that they ignore: the proliferation generated by their work of purification (Latour 2001, p. 34). An important process

---

1. (Grinde & Johansen 1995)

for re-situating modern thinking involves dialogue. Open, situated dialogue is a place of encounter for moderns and amoderns. It is an opportunity for people to meet, communicate, translate and be transformed. This is currently happening at water management negotiation tables everywhere.

FALSE CHOICE

Moderns insist on a particular historical pathway which has led to the development of categories such as nature, nation and economy, and with repetition these categories are understood by moderns to be so real as to 'exist prior to any such representation' (Mitchell 2000, p. 19). They become implicitly and explicitly regarded by moderns as universals based in reason that is culturally neutral and omnipotent. Thus, the moderns presume their governing principles are impartial truths. This modern thinking created uniform citizenship as the basis of modern governance — in rhetoric but not in practice (Tully 2004a, p. 9).

Such modern concepts and cognitive failures are part of state institutional practices that make it difficult to acknowledge other political presences. For example, the expansion of the forestry industry in British Columbia, Canada, was only possible through the representative force of nature and nation (Braun 2002, p. 8). In colonial forestry reports nature and nation were made to coincide, 'the former produced rhetorically as the prehistory of the latter' (ibid., p. 66). The sovereign territorial claims of the Canadian state could then become the starting point for governance, and the historical–geographic constitutions of the forest lands were explicitly ignored. The temperate rainforests became at once 'deterritorialised and re-territorialised'. These erasures continue in contemporary Canada. In the late-twentieth century dispute over the same forest, the forest company employed concepts of the nation, national interests, and the public to speak with authority within this construction. The First Nations peoples could only enter this constructed 'public' as a special-interest group (Braun 2002, pp. 41–3).

Political scientist James Tully argues that when Indigenous people engage with settler states the limited choices they face are a result of the way state institutions and languages have been framed (Tully 2004a, p. 56). What appears to be a fraught choice — between co-option or marginalisation — is actually an outcome of institutional design because the settler state does not have flexibility for cultural diversity incorporated into its processes and approaches. James Tully has researched how state institutions used to have the capacity to accommodate cultural diversity through three principles: mutual recognition, consent and continuity. That is, agreement by both parties on a form of mutual recognition; the consent of the peoples in relation to decisions that affect them; and the continuity of peoples' laws and customs with the new political associations. Subsequently, in the evolution of state institutions, these earlier values have been lost to arguments for uniform political and legal order (Tully 2004a, pp. 116–29). Examples of a fraught choice in relation to institutional design are evident in western environmental programs that assume that 'nature' is separate and subordinate to 'humans'. When the MLDRIN delegates are invited to participate in such government environmental programs, they have to consider how their participation legitimises those programs and co-opts their own approaches. They also lose the negotiating tool of not participating, and must accept marginalisation within the dominant uniform framework (Tully 2004a, pp. 57–8).

This situation is exemplified by the Yorta Yorta's participation in the Barmah–Millewa Forum. This forum used to advice the Murray–Darling Basin Commission about forest and water management of the wetlands, which are at the heart of Yorta Yorta country. Frustrated about being marginalised and ignored in its procedures, the Yorta Yorta chose to leave the forum rather than legitimise its decisions through participation. As Yorta Yorta Elder Henry Atkinson said (at a MLDRIN joint-meeting with the Community Advisory Council, held in Wagga Wagga in July 2005), 'We had a chair at the table, but we had to keep our voice outside'. The forum was set up to represent the 'wide ranging interests of the forest' (Anon. n.d). By leaving the forum, the Yorta Yorta made the point that this representative forum should represent their voice if they want their participation, and not just co-opt their participation within the forum's agenda. The Barmah–Millewa Forum has since been disbanded. Its purpose was challenged by the new consultative structures for the Barmah–Millewa forest as an icon site under The Living Murray, and by the conflict with the Yorta Yorta.

Further downstream, the Ngarrindjeri have also had an energy-draining experience with modern environmental management. In this case it was the Government of South Australia's plans for managing the 'wise use' of Ramsar wetlands the Coorong, and lakes Alexandrina and Albert (areas declared a Wetland of International Importance in 1985 under the Ramsar Convention (Hemming 2005, p. 8)). The Ngarrindjeri were provided with funding, time and expertise by the Ramsar planning team to mobilise their involvement. However, the report the Ngarrindjeri prepared was not included in the planning outcomes, not even as an appendix to the management plan (Hemming et al. 2007, p. 21). Ngarrindjeri participation became 'subject to South Australian government policy relating to the resolution of native title claims' (DEH 2000, p. 2). That is, the South Australian

Government decided to withdraw from the engagement until the Ngarrindjeri native title application was determined. Native title is indeed unsettling the moderns. Native title holders are recognised as having their own systems of law and governance; however, the federal government has responded by trying to bring native title within the property law system (Strelein 2007).

For governments, Indigenous peoples and their knowledge have to be brought into line with the 'universal' framework of modern political–legal institutions. Geographers Richie Howitt and Sandie Suchet-Pearson have described Indigenous peoples' participation in projects managed by government (Howitt & Suchet-Pearson 2006, p. 32; my emphasis):

> Developmentalist and conservation projects have long sought to *discipline* indigenous peoples and their domains to bring them within the compass of mainstream management practices.

When Indigenous peoples articulate knowledge that does not fit within the project framework, it is the knowledge that is deemed to be inappropriate or deficient, not the project framework. As the value of Indigenous ecological knowledge is now increasingly acknowledged in Australia, the environmental managers struggle to insert Indigenous knowledge as 'additional data' within their scientific knowledge framework. The impracticalities of this process can result in Indigenous knowledge being criticised by the environmental managers for being hard to understand or access (Hemming et al. 2007, p. 116), without a concomitant reflection about the role of modern knowledge in setting the terms of the translation process.

Moderns make 'circular arguments' (Howitt & Suchet-Pearson 2006, p. 323). As Debbie Rose has described, the moderns exist within a 'hall of mirrors' where they engage in monologue with their own reflection (Rose 1999, p. 177). Universal reason enables moderns to take universal action which they do not need to examine for cultural bias. Their belief that their knowledge transcends culture means they do not need to consider whether their initiatives include equitable treatment of cultural diversity (Tully 2004b). The modern conceptualisation of 'sameness' within the community is exemplified by the stakeholder or interest group model in environmental management, within which it is presumed that all people can be accommodated (see, for example, Aslin & Brown 2004). The approach denies, or is blind to, Indigenous peoples' governance relationships with country. Natural resource managers need to recognise that Indigenous people have their own methods for determining what is happening, why and what the appropriate response is (Palmer 2006, p. 35). Indigenous peoples' knowledge is not simply data about the habits and habitats of plants and animals. For traditional owners, knowledge about country is part of a cosmology within which they hold ethical relationships and responsibilities for the life of country.

Despite their efforts for uniformity, modern nations are populated by diverse cultures and must manage the cultural recognition claims that are made within their borders. Nations do not represent a single isolated culture. Rather, cultures overlap geographically within and between nation states, are interdependent in their formation and identity, with complex histories of interaction and negotiation (Tully 2004a, pp. 5, 10–11). This is expressed by the term 'intercultural', which academics use to reject notions of cultures as exclusively bounded, self-defining and self-reproducing domains (Martin 2003, p. 4; see also Hinkson & Smith 2005 and Eriksen 2001). Cultural groups who are minorities within nation states and wish to negotiate with nation states must manage the modern preference for uniformity.

Dialogue, not monologue, is valuable for engaging this cultural diversity. Governing principles are not inflexible universal truths that people either choose to support or deny. Moreover, Indigenous peoples' critiques of these governing principles are not just claims for greater involvement, but are the positive expression of their own political identity. When negotiations occur between parties with different traditions of governance, dialogue is a valuable tool for understanding. Both sides of the negotiation may be communicating in different languages, but through listening and translating, a level of understanding can be reached (Tully 2004a, pp. 110, 132–3). Further, through dialogue a sense of partnership can be achieved (ibid., p.100).

Because there is such diverse knowledge involved in engagements between Indigenous people and governments, Debbie Rose argues that settlers should participate in an 'open' dialogue with Indigenous people (Rose 2004b, p. 22); that is, be vulnerable so that one's own ground can be destabilised as part of the dialogue. Both sides of the negotiation have to examine their assumptions so that blind spots, stereotypes and prejudices are addressed and overcome. This will create the space for a more respectful dialogue. Dialogue in itself is not the end point for the resolution to cultural diversity claims. The terms of dialogue have to be examined for inequitable power relationships between the parties that meet (Tully 2004b, pp. 93, 94). Moreover, dialogue about culture is an ongoing process in an intercultural context.

Returning to Donna Haraway's argument for situated knowledge, Richie Howitt and Sandie Suchet-Pearson call for 'situated engagement' as an antidote to the power of universals (Howitt & Suchet-Pearson 2006, p. 332). Similarly, I argue for this knowledge to be engaged in dialogue that emphasises the exchange of knowledge from particular places. In situated dialogue each party can talk to each other cognisant of the assumptions they bring from their particular experiences. However, I wish to hold onto what Anna Tsing calls 'sticky universals' (universals that 'stick' and 'grip' to places); that is, universals that have meaning to places, but also have meaning in broader forums because universals can play a powerful role in knowledge that 'moves' (Tsing 2005). We do not have to throw out the Enlightenment but we must be cognisant of the assumptions behind the Enlightenment principles.

THE LANGUAGE IN NEGOTIATION

In negotiations between Indigenous peoples and governments, missed meanings are often prevalent: the parties often do not understand the meaning behind the language being used (Strelein 2002–03). The languages and terms used by parties to communicate at the negotiation table are invested with their intellectual frameworks. Indigenous people can make cultural recognition claims in their own languages, but in Australia in meetings with governments the only language that is spoken is English. Thus the work by Indigenous people to create the space to 'be' Indigenous must be explicitly translated. For the majority of Indigenous peoples in the Murray–Darling Basin, English is their first language, yet translation is still necessary because they have inherited a different knowledge tradition which they express in English and Aboriginal English. When the same words are used by different parties with different meanings, missed communication is the result.

An example is the term 'natural resource management'. This is a relatively recent addition to the English language but is already in constant use in government discussions about the inland rivers and ecological restoration. Yet the exact meaning of 'natural resource management' is not defined in the glossary of the Murray–Darling Basin Commission's annual report or in the glossaries in The Living Murray documents. Its meaning is assumed to be self-evident. A definition can be found in the *Natural Heritage Trust of Australia Act 1997* (Cwlth) (s.17) and the *Natural Resources Management (Financial Assistance) Act 1992* (Cwlth) (s.4). In both these Acts, 'natural resources management' is defined as:

(a) any activity relating to the management of the use, development or conservation of one or more of the following natural resources:
(i) soil;
(ii) water;
(iii) vegetation; or
(b) any activity relating to the management of the use, development or conservation of any other natural resources for the purposes of an activity mentioned in paragraph (a).

Thus, natural resources management is implicitly a human activity. The humans include farmers, irrigators, forest operators, public utility managers, local government officers, catchment authority officers, regional natural resource management bodies, community volunteers and many others (Australian Government 2007). This definition is consistent with modern thinking that constructs nature as separate and subordinate to humans, where humans intervene and exert control over a static nature. Further, the humans are responsible for 'progressing it' from an 'original, wild state to a developed, civilised and domesticated state' (Howitt & Suchet-Pearson 2006, p. 324).

In a break during a MLDRIN meeting in Echuca in March 2004, Mutti Mutti Elder Jeanette Crew and Yorta Yorta woman Monica Morgan talked together about when they first realised that natural resource management did not mean 'caring for country'. The governments' assumptions behind the term natural resource management were not immediately apparent to either woman. Jeanette and Monica both held implicit assumptions about looking after country, which they came to realise were culture-specific when they experienced miscommunication at the negotiation table.

In this conversation, Jeanette recalled a water management meeting in which she asked other participants whether they thought natural resource management was a good thing. The response was silence. Examining the merit of the term was an unexpected question. Monica recalled how when she went to work in Canberra for the Murray–Darling Basin Commission she learnt that natural resource management had a very different meaning to caring for country. She (re)interpreted it as 'controlling resources for the irrigators'. This is supported by Libby Robin, who analysed how the term 'natural resources' transforms nature (Robin 2007, p. 186):

> when elements of nature are selected out as 'natural resources' they are no longer natural. Their local ecological context is replaced by a global economic one. Reading a landscape through the lens of an unnatural economy changes the way people see, understand and identify with nature.

In her interview with me Jeanette spoke about her traditional country as a place alive with mythical creatures, including the Bunyip who lives in the deep pool in the river, and the *murrupmaginnie* and the *nyutha* who live in the hollows of trees. As Jeanette said, the hollow trees:

> were homes for those creatures, so you always had to have them. And nowadays you have to have them because they are habitat for native fauna. [20 July 2004]

As Jeanette said that last sentence her voice flattened. Instead of a natural resource management which only records, values and cares for an external world of resources, Jeanette envisions a natural resource management in her country which includes the other creatures she grew up with. Jeanette can see country as a place full of energy, where information is exchanged between humans, plants, animals, creatures and ancestral beings. In this natural resource management the scope of human activities is enmeshed with the agency of many extra-human others. As Ben Smith also described through his work with Indigenous people, this 'management' is not human action *upon* the world, but 'an underlying field of connectivity which is reproduced and maintained, in part, through the human agency it provides for' (Smith forthcoming, p. 9). This management repositions human activity within webs of ecological life, and acknowledges co-dependence as inherent to these relationships.

The failure of governments to acknowledge and address Indigenous knowledge as part of the design of their partnerships with Indigenous people has caused conflict in natural resource management. These conflicts are not simply about competing vested interests in shared spaces, but are a sign of deep ontological schisms (Howitt 2001, p. 59). Using the example of co-managed national parks, Richie Howitt argues that most co-management arrangements are structured in a way that incorporates Indigenous people into the dominant paradigms that define resources and their management (ibid., p. 374). Instead of being able to bring their own management to the negotiation table, the traditional owners have to expend energy in convincing the other side to examine their own assumptions. This diverts attention away from the intention of the joint-management — to reach agreed approaches — and undermines the contribution that Indigenous people bring to negotiations with government about the management of country.

Natural resource management is an example of a recently coined term that has developed in close connection with modern conceptions of nature. It is a consequence of hyper-separation. Across the world, nature is grasped in consumptive terms by the moderns. This is true both in relation to consuming resources without any consideration for the future, *and* sustainably managing resources for consumption in response to scarcity. As a new term in a modern framework its interpretation is very specific.

One word that carries considerable influence in the context of Indigenous people negotiating their rights with government is 'tradition'. The High Court has discussed the meaning of tradition in relation to native title (*Members of the Yorta Yorta Aboriginal Community v Victoria* (2002) 77 ALJR 356, para. 46):

> A traditional law or custom is one which has been passed from generation to generation of a society, usually by word of mouth and common practice.

For traditional owners, tradition is central to their identity; they identify themselves as a distinct group of Indigenous people: the traditional owners of country. Carrying on their traditions is an important part of their contemporary identity. These traditions include the myriad of interpretations and adaptations, as well as adoptions that occur over time either through personal choice or as part of engaging with other cultures. However, the theory of progressive reason equates changing traditions with loss for premoderns. From the modern perspective, cultural change is constructed as antithetical to the continuance of Indigenous peoples' traditions.

The recognition of native title has highlighted these different interpretations of tradition. In native title, Indigenous people can have their rights to country recognised only if they are able to demonstrate to the courts that they have traditional laws and customs that have been maintained since colonisation (*Native Title Act 1993* (Cwlth) s.223). This is what the Yorta Yorta people had to do in their native title trial (*Yorta Yorta Aboriginal Community v the State of Victoria and Ors* (1998 unreported)). The Yorta Yorta argued that they were responsible for their country because they are the traditional owners of that country, which was created for them and which supports their life (Atkinson 2004). The Yorta Yorta talked about their aspirations for and participation in cultural heritage and national resource management as part of their identity as traditional owners (Muir & Morgan 2002, p. 5).

However, for the Yorta Yorta people their native title determination hinged on a different interpretation of 'tradition' by Justice Olney. In 1881, the Yorta Yorta people had petitioned the governor of New South Wales for some of their traditional country so that they could become independent farmers. This petition had been presented to the Federal Court by the Yorta Yorta as evidence of their ongoing connection to country as identifiable peoples. However, Federal Court Justice Olney viewed the adoption of commercial farming by the Yorta Yorta as antithetical to their status as traditional owners (Strelein 2006, p. 85). Indeed, Olney determined that by pursuing commercial farming the Yorta Yorta had 'abandoned' their native title. Olney had wanted something else from the trial: contemporary evidence of initiation or other religious and/or ceremonial practices. Olney also argued that environmental conservation was not (Yorta Yorta) culture (paras 126, 128).

On appeal, the High Court substantiated Olney's interpretation of tradition by methodically defining the meaning of 'tradition' under native title law, as the laws and customs that existed before the British asserted sovereignty. Further, that (paras 46–7):

any later attempt to revive adherence to the tenets of that former system cannot and will not reconstitute the traditional laws and customs out of which rights and interests must spring if they are to fall within the definition of native title.

While the High Court rejected Justice Olney's 'abandonment' thesis, the High Court was not able to recognise the continuity of connectivities and emergence of new connectivities and engagement with new knowledge as constituting 'tradition'.

In his judgment, Justice Olney did not find in favour of the Yorta Yorta people and determined that the Yorta Yorta's traditional laws and customs had been 'washed away by the tide of history' (para. 19; this phrase was used repeatedly in the determination, see paras 3, 126, 129). That is, the extinguishment of their native title rights and interests was an unfortunate but inevitable part of Indigenous peoples' path through history. The 'tide of history' is an assertion of the absolute primacy of the colonial narrative, which sterilises the violence of colonisation as the tidal movement of water (Ritter 2004, pp. 117–19). This event reconstructs the 'national failure' by successive governments to address the injustices of the dispossession of Indigenous people into a 'local failure' of the Indigenous people who failed to be authentic (Povinelli 2002, p. 56).

Yorta Yorta Elder Henry Atkinson spoke about the impact of the Yorta Yorta native title decision on his life (Arnold 2006):

> It made me feel like I didn't exist. It feels like you've lost your identity. Who are you? Where do you really come from? How can you identify yourself as a person?

In native title, theories around tradition and change can be used either for the oppression or liberation of Indigenous people. Justice Olney relied on hyper-separations and the theory of progressive reason so as to deny native title, in doing so denying connectivity and, as Henry says, negating identity. Conversely, in the High Court's 1992 *Mabo* decision (*Mabo v Queensland [No.2]* (1992) 175 CLR 1), Justice Brennan identified that the theory of hierarchical civilisations was racially discriminatory in order to recognise native title. The *Mabo* decision has become popularised as the overturning of the doctrine of *terra nullius*, a doctrine closely linked to moderns' theory of hierarchical civilisations. As Justice Brennan wrote in his judgment (para. 39):

> The theory [*terra nullius*] that the indigenous inhabitants of a 'settled' colony had no proprietary interest in the land … depended on a discriminatory denigration of indigenous inhabitants, their social organization and customs … the basis of the theory is false in fact and unacceptable in our society.

In *Mabo* the Justices drew on contemporary understandings about race and racial discrimination, developed in part by international rights discourses, to overturn the theory of hierarchical civilisations and recognise native title.

Formulas of identity have always been a part of the management of Indigenous people by governments (Dodson 2003, p. 34). Indeed, governments can use the recognition *and* the denial of cultural difference for the purposes of discrimination (Macklem 1991, p. 392). This has been the experience for most of the Elders in MLDRIN. They grew up on Aboriginal missions, which were a response by government to the management of Aboriginal people segregated from the rest of society. On the missions Aboriginal people needed permission from the mission manager for the most fundamental of their activities: to work, move house and marry. Today, these same people battle for the recognition of their rights as Indigenous people and traditional owners. The life experiences of the MLDRIN delegates show how governments have used cultural difference first to exclude, and then negate, that difference to deny rights. The violence to the Yorta Yorta people's identity resulting from the native title decision is a cautionary tale about what happens when oppositional dualisms are part of an exercise in power and law-making over the identity of Indigenous peoples. In native title, the hyper-separation of tradition is used in the courtroom by one culture sitting in judgment over another (Strelein 2000).

FRAMING INDIGENOUS PEOPLES' SUBJECTIVITY

When a cultural group applies to government for recognition, both the people claiming the rights and the government are forced to bring into being a collective subject: the 'rights holders'. Determining the boundaries of this subject position happens through a process which maps rights into identities and vice versa. The MLDRIN delegates are very aware of this construction: their Indigenous identity is packaged and a performance of identity is required in negotiations with government. For example, Ngarrindjeri Elder Matt Rigney deliberately articulates his identity to express what is meaningful to him and to challenge the boundaries of modern thinking. I asked him about the tidal fish traps at the Coorong that he had shown me on an earlier visit:

> The fish traps, they were built by our old people, often we refer to them as our ancient ones. We are proud of them. I think it is a form of technical innovation, I suppose you could call it. I suppose they got tired of standing on the shores, like a rock in the water, waiting for a fish to come by and spear it. They thought they would become more commercial, so they built fish traps. [24 July 2004]

4. SETTING THE NEGOTIATION TABLE

Together we laughed about the commercial reference. He then quickly followed on by describing the traps as 'mass production'. Matt deliberately shows he knows what is required to even begin to have his economic rights recognised within modern thinking. When we stopped laughing, Matt became more serious and talked to me about the fish traps in relation to cultural change. Matt moved on from deliberately fitting in with modern thinking to critiquing its philosophical shortcomings:

> But it is an important thing to understand how our culture has developed. Often people see culture as a static thing, and not as a living thing, and we all know that culture is a living, continually growing philosophy … It reshapes and reforms itself every minute of the day. And our people used that technology to sustain their life in a more quality way. [24 July 2004]

Matt makes the point that if pre-colonial Aboriginal culture was a part of a continually changing cultural process and practice, then modern thinking should be able to acknowledge this dynamism and to acknowledge the dynamism of today's Aboriginal culture.

The fish traps are made by placing stones together in a circle in the shallows, and the fish are caught when the tide goes out. Their stone rings can be spotted breaking the water's surface along the shoreline of the Coorong. Matt said the fish traps were not being used anymore because they were not in very good condition. Matt talked to me about rebuilding them:

> They have been neglected because we have not had … we've been unsure about how government, especially the National Parks who look after the Coorong, have management control over it, how they would view us going into the Coorong and building [with] rocks. Placing rocks on top of each other. So we haven't done that, because we didn't want to disturb anybody or get into a fight with anybody, and now we are saying, 'Well, now, we're going to start going back and practicing our traditions', whether they like it or not. They can take us to court if they wish to stop us. And we'll see what happens in the system. [24 July 2004]

Being a traditional owner and being responsible for the fish traps in the Coorong involves multifarious interactions with governments, with different opportunities and challenges. If the South Australian National Parks and Reserves people prefer that the stones that form the fish traps are not 'disturbed', Indigenous people can seek to protect their rights in court, and they will have a chance at litigated success as catching fish this way is so evidently connected to pre-colonial traditions.

Rebuilding the fish traps is a part of the continual process of creating and recreating Indigenous cultural identity. The government might support this

Ngarrindjeri activity or it may not. Either way, exercising contemporary Indigenous identity involves negotiating with governments, because Indigenous people believe they have responsibilities for places that governments believe are managed under state and territory structures.

The relationship between Aboriginal identity and modern thinking is also discussed by Steven Ross, the executive officer of MLDRIN. Steven related to me how his childhood experiences are framed within broader government and academic discourses. We were in Deniliquin by the Edward River at a spot called MacLeans Beach, when Steven said:

> It is funny, when you go to uni and you go into the public service and hear all this kind of — particularly in Aboriginal affairs and natural resource management — there are the catchphrases right there … around socioeconomic status, cultural heritage and all that. But when you are a kid you don't know any of that stuff, you just kind of do it. It is nice that we are here at MacLeans Beach because we always used to come down here … and collect the mussels and take them home and eat them. And you can go almost anywhere along this river and you could get yabbies. It is weird now attaching those titles to it. It becomes kind of formalised and analysed. It is strange it has become part of the dialogue now I think. Where, actually, blackfellas just do it. So that schism's a bit weird. [28 July 2004]

For Steven, life can be the experience of simply living or it can be intellectualised through the catchphrases that are used in government policies. Steven's consciousness of this is a very explicit comment on the ways Indigenous subjectivity is framed and split by other knowledge frameworks and government practices. Steven is expressing the consciousness of being 'other' in a field of differential power and loaded discourse.

Identities are complex and include elaborate dynamisms of attractions and aversions, yet they are packaged in a bounded, consistent way in government law and policy, and consequently in dialogue with government (see, for example, Smith 2003). This abstraction of identity belies the diversity of individual life experiences, and can confound Indigenous people who are expected to meet the examples that are set out for them by law and policy processes. Anthropologist Elizabeth Povinelli wonders about the constraints of these identity constructs, and asks the rhetorical question as to whether an Indigenous person who watches the television show *Days of Our Lives* can be entitled to valuable land on the basis of their traditional beliefs and practices (Povinelli 2002, p. 60). The pressure to be consistent about identity can lead to a complicated existence of handling dual identities: declaring the wholeness and boundedness of your identity at the negotiating table, and then outside the room adjusting back to the daily reality of negotiating complex intercultural life, in all its glamour and ordinariness. In response to narrow understandings about culture and identity, philosopher Paul Hountondji succinctly argued that 'culture is not only a heritage, it is a project' (cited in Sahlins 1999, p. xxi). Culture is about process and practice. The *way* people change their cultures can be understood as culture, as the changes work in reference to the meanings and practices of that culture.

Marshall Sahlins has argued that the theory of progressive reason is smashed by Indigenous people creating their own versions of modernity (Sahlins 1999). Instead of disappearing as predicted, Indigenous people have incorporated and adapted different tools of modernity for their own purposes. More than this, what Indigenous people are doing is not limited to versions of modernity — they are establishing their own narratives of resilience and identity. Far from being positioned as remote in time and space, Matt and Steven show how they use the language and tools of the moderns as they seek to make their own space in their negotiations with government frameworks.

IDENTITY, SURVIVAL AND RIGHTS

All the traditional owners I interviewed have a strong appreciation for continuing cultural practices from the past, which they call their traditions, and they wish to protect these traditions. Despite the prejudicial use of tradition by governments and courts, the traditional owners continue to identify with the term. Indeed, they counter-theorise how tradition and cultural change go together with their experience of surviving colonialism. They also appeal to the modern language of rights to protect their traditions.

Yorta Yorta man Lee Joachim talks about how he understands the past as a part of the future of the Yorta Yorta people:

> [We need to] show and respect our cultural practice that we have practised over tens of thousands of years and is still here today. To remove ourselves, to decolonise ourselves in the way of life that has been forced upon us, and to get back to our true grassroots levels of where the Elders were guiding us, the spirits were guiding the Elders, and we as the young folk were going through initiation stages to become a part of the bigger picture and to become the Elders of the future. We don't have to go back to lap-laps and going out there and living life as it was back 10 000 years ago. We can still marry in today with what culture there is today, but we've got to give true respect back to our Elders. [25 June 2004]

Lee is interested here in processes which will strengthen Yorta Yorta knowledge traditions and lines of authority. He is not interested in going back to how things were before the British asserted sovereignty. Lee wants to develop dynamic processes that strengthen relationships between the people, the ancestors and future generations.

Mutti Mutti Elder Mary Pappin spoke to me about how Indigenous people managed the disruptions to their lives caused by colonisation. She called it a storm:

> We can ride out the storm, hopefully. I believe our ancestors rode out many storms and I think we can do it too. If we keep our cultural heritage intact for our family groups, for our next of kin, for Aboriginal Australians, if we can keep the cultural heritage intact and alive and well, we can ride out anything. [22 July 2004]

Mary describes cultural heritage as the vehicle for their continuing survival as identifiable peoples. Mary's phrase keeping the 'cultural heritage intact' might imply that culture is unchanging and bounded, but I interpret it as her desire to pass on cultural heritage from the past that gives rise to the language of protection. Mary relates heritage to the adaptations that Indigenous peoples have made since colonisation:

> Because we have adapted doesn't mean to say that we are one of them, it doesn't mean to say that we have to give up our cultural heritage. We can have them side by side because 60 000 years of cultural heritage and people belonging to a place did not change the people to make them something else. [22 July 2004]

Mary strongly identifies with a heritage of people who can survive in harsh arid environments, and she talked with me about the survival of her people when Lake Mungo dried up 20 000 years ago. By using these time frames and experiences as a reference point, Mary discounts the impact of colonisation. Indigenous people can point to their long history of occupation to assert their identity as First Nations; it is undeniable that they were here first and have been here for a very long time.

Mary spoke to me about her responsibility as an Elder to look after important sites, bush tucker and medicines, and to pass on her knowledge to the next generation:

> Because otherwise people will forget who they are, what their rights are. Our born right is [to be] kept intact because even though we may not have ownership of the country, the land does belong to its people and we are the people from that land. [22 July 2004]

The usually drowned Yellowin Bay emerges when the water level drops at the Blowering Dam, Tumut, New South Wales.

These flat plains in south-central New South Wales,
Wiradjuri country, stretch for hundreds of kilometres,
and are pastoral lands and emu habitat.

The beauty of a dead river red gum forest, Lake Mulwala on the Murray, New South Wales, upstream from the Barmah-Millewa forest.

Young river red gums in the Werai forest, near Deniliquin in New South Wales, Wamba Wamba country.

The Murray River at Redcliff, south of Mildura in Victoria.

Fivebough Wetland near Leeton, New South Wales, Wiradjuri country.

The Murray River at Wentworth, New South Wales, is a popular holiday destination.

Carpark Reconciliation. A mural painted by volunteer university arts students in 2000, Meningie, South Australia, Ngarrindjeri country.

A crack in a midden next to the Murray River at Redcliff, Victoria, shows layers of mussel shells embedded in the red dirt.

4. SETTING THE NEGOTIATION TABLE

Mary links tradition, knowledge, rights, culture and land together in what she calls a story of survival. Mary presents these as indisputable facts: born rights. That is, rights as something you *are* rather than something you *have*.

Significantly, universal rights language has provided tools of engagement for Indigenous peoples who are being discriminated against in their nation states. The idea of universal human rights derives from the events surrounding the Holocaust in mid-twentieth century Europe. This state-led mass murder starkly revealed how a government grossly failed to acknowledge the human rights of its own citizens. The Holocaust also showed the need for rights that are enforced by the whole of humanity so that people are not victimised by their governments. The drafting of international United Nations human rights documents is intended to build a balance between citizens and the authority of sovereign states. Rights frameworks create grounds for citizens to disobey the legal yet immoral orders of their rulers, thus returning agency to individuals (Ignatieff 2003, pp. 4–5). As Robert Williams, a legal academic and a member of the Lumbee Indian Tribe of North Carolina, has argued (Williams 1990, p. 701):

> The discourse of international human rights has enabled indigenous peoples to understand and express their oppression in terms that are meaningful to them and their oppressors.

Indigenous people use the rhetoric of the western rights discourse to challenge institutions internally with their own logic (Crenshaw, cited in Williams 1990, p. 701). This is part of the politics of negotiation. Appealing to universal rights frameworks usually means appealing to governments for protections, exemptions or inclusions in legislation and policy. The MLDRIN delegates draw on rights language as part of their negotiations with water bureaucracies. Indeed, they asked academics to write a discussion paper about Indigenous rights to water (Morgan et al. 2004; I was a co-author of that publication). Yorta Yorta man Lee Joachim talked about how the protection of Indigenous rights is a part of his cultural survival:

> I would just like to see some rights be issued back to our people to ensure that our culture remains alive and we as a people connected to our country remain alive along with it. [25 June 2004]

When Lee talks about rights he uses the phrase 'issued back', which is a reference to the rights talk of historical injustices. For culture to remain alive, and to be carried between generations and into the future, rights are needed to protect it. In this way culture is seen to be flowing through time. Mary and Lee want to protect this flow, and influence and recharge it.

Establishing the recognition of certain rights is important because constantly negotiating one's identity with government is exhausting and confronting work. With recognised rights to country, the traditional owners can create some uncontested mental and physical space to 'be' Indigenous, and have a break from negotiating and performing to modern knowledge. This includes going fishing without having to apply for a fishing licence or, as Henry Atkinson puts it, 'without the need for a piece of paper to say what I can or cannot do' (Atkinson 2004, p. 25).

However, Indigenous peoples' rights claims within settler states draw them more deeply into intercultural spaces, even as they claim their own space to be Indigenous. As Ben Smith has observed, Indigenous people often express their identity in oppositional terms to the mainstream, but it is within the social and cultural horizon of the mainstream that Indigenous people (and non-Indigenous people) name, intensify and ratify Indigenous identities (Smith 2006, p. 228). Despite this intercultural intensification, the traditional owners continue to identify and articulate a separate identity which they perceive as profoundly central to their work to restore the health of the river country. They do not wish to 'discipline' their knowledge to conform to the mainstream, but, as Lee said, 'marry in' their traditions with contemporary Australian society. This intercultural process is about agreement not imposition.

IN DIALOGUE WITH UNIVERSALS

The way universals are presented by moderns as irrefutable rationality means dialogue with the moderns is often more like monologue. Mary and Lee show us how they work to counter modern thinking whilst seeking the protection of universal rights frameworks. Mary and Lee use universal rights language to create the space for dialogue, because rights talk is meaningful for both Indigenous people and moderns.

Anna Tsing has analysed the different possibilities of universals. It is the capacity of universals to move and translate across cultural difference that interests her. Whether they underlie or transcend cultural difference, universals form bridges between different knowledges. This is important in culturally diverse societies. Anna Tsing has argued that this movement is desirable for all parties, elites and the excluded alike. In negotiations, all parties wish to translate and transform the engagement by appealing to universal meanings (Tsing 2005, pp. 7, 9).

According to Anna Tsing, universals should be grasped 'not as truths or lies but as sticky engagements' (Tsing 2005, p. 6). She has argued that specific knowledge can give 'grip' to universals, grip that is critical for universals to find purchase in everyday places in order for the concepts to travel. Indeed, the biases and prejudices within many universals situate these knowledges somewhere; that is, in the west. Anna Tsing's approach to universals is in line with Bruno Latour's arguments: that we do not have to abandon modern knowledge just because modernity has been so powerful and destructive (Latour 2001; see also Tully 2004a). Instead of the dead end that is the subjectivities of postmodernism, we need to grasp that which connects us and can be exchanged meaningfully in dialogue with each other. Understanding modern knowledge is useful if we can identify and avoid the mistaken beliefs of hyper-separation (such as nature/culture, tradition/change), and avoid blindness to the effects of this hyper-separation (such as ecological destruction and the violence to Indigenous peoples' identity). We should use universals to support, rather than suppress, communication.

Ngarrindjeri Elder Richard Hunter talked to me about the way dialogue was important in his local community after native title was recognised in *Mabo*. Local non-Indigenous people were worried about losing their shops, their land and their crops to native title claims. Misunderstandings and misapprehension about native title had built up so much that the local traditional owners called public meetings for the community to talk about it together. As Richard recalled these meetings, he said:

> We just wanted, say, the right to come through your land to see what's on our sand, our burial sites are there. Our culture is out there, as well as yours, the Europeans. We've got to share it, as simple as that. If you want to get technical about it, well we were here before you. And yet you've never acknowledged the Indigenous people that … We're both trying to survive in the one land. [23 July 2004]

Devising formulas to close down dialogue will not transcend the fact that this is shared country. Rather, strategies to communicate, co-exist and understand each other are needed. This is dialogue that is situated in place. The *Mabo* decision had been worked out in the Torres Strait and then applied to the Australian mainland. But the mainland had not been engaged in the process of understanding what native title might mean in their locality. The Ngarrindjeri initiated their own local dialogue to talk through these matters with their neighbours.

Because of the different knowledge traditions at play, dialogue can be tricky and parties to negotiation can abuse their different sources of power. Indigenous people also have to seek paths of communication, and must be cautious with rights talk. In negotiations, rights can be used as trump cards where rights are presented as non-negotiable absolutes, and rights talk can close down dialogue exactly when dialogue is needed to negotiate the rights. Everyone has rights hanging in the balance at negotiation tables (Ignatieff 2003, p. 20; Brown 2003, pp. 226, 231). But we should not be quick to position rights as oppositional. Healthy river systems support a range of rights for a range of human and non-human actors. As Wiradjuri Elder Ramsay Freeman said:

> We don't only fight for our rights, we fight for everyone's rights. Not only the Indigenous, but everyone. Non-Indigenous and all. We're fighting for the water, which we think is valuable to everyone ... None of us can survive without water. [27 June 2004]

The many tensions and fractious understandings encumbered in dialogue between Indigenous peoples and governments about Indigenous rights and identity were revealed in the bitter contest known simply as 'Hindmarsh'. This conflict became highly politicised, touching a raw nerve in the Australian psyche about who can say what and how. The dispute pitted local religious beliefs stated by a group of Ngarrindjeri people against marina developers who were focused on building a bridge to connect Kumerangk (Hindmarsh Island) to the small town of Goolwa, just behind the Murray Mouth in South Australia. The Ngarrindjeri people argued that the bridge would desecrate a sacred place in their country. The developers wanted the bridge to provide land access to the marina on the island.

The bitter Hindmarsh dispute is better described as a 'cataclysmic encounter' than a dialogue (Marcus 2004, p. 340). The conflict involved court cases and appeals under state and federal heritage legislation, a royal commission, a High Court appeal, and a year-long trial in the Federal Court in 2001 (which was the biggest of 15 defamation cases brought by the marina developers) (Ogle 2002, p. 16). The dispute was played out in the media and in parliaments, and people watching it on television took their own positions about the validity of the local religious beliefs of a group of Ngarrindjeri. In the final defamation case, Justice von Doussa determined that the women's religious beliefs were not fabricated, and was very critical of the bridge proponents (*Chapman v Luminis Pty Ltd (No 5)* [2001]FCA 1106).

Debbie Rose asks how can Aboriginal people 'expect to "enjoy" their culture' when their laws and lives are 'scrutinised, debased and debated in a form of public vivisection' (Rose 2001, pp. 115–16). She questions the role of scholars in these engagements and wonders whether our engagement 'simply serves to sustain a pretence of open debate in a plural society, while we hover at the periphery of the destruction that continues around us' (ibid., p. 116). Instead of scrutinising the authenticity of Indigenous culture, scholars must ensure Indigenous people have the intellectual space to express themselves in Australian society. Otherwise the debate is a fiction.

Since Hindmarsh, the Ngarrindjeri and the local government have been working together on a different dialogue. In 2002, a 'Sincere expression of sorrow and apology to the Ngarrindjeri People' was signed by the mayor of Alexandrina Council and witnessed by three Ngarrindjeri Elders and the council chief executive (Alexandrina Council 2002). This document acknowledged and expressed regret for the suffering and injustices that Ngarrindjeri people had experienced over 166 years of colonisation. The council also explicitly recognised the Ngarrindjeri people as the traditional owners, their unique culture, and 'your right to determine your future' (ibid.). It is a statement of reconciliation that acknowledges the past and lays the foundations for agreement-making into the future. The Ngarrindjeri people now work closely with the council on issues of concern to the traditional owners, including development applications that are next to the Murray River. They have also begun working towards a formal agreement with the Ramsar managers of the Coorong (Hemming et al. 2007, p. 225).

Further upstream, in 2004 the Yorta Yorta negotiated a Yorta Yorta Cooperative Land Management Agreement over the Barmah–Millewa forest with the Victorian Government. Henry Atkinson has written that this agreement-making experience was based in recognition, mutual respect and shared goals (Atkinson 2004, p. 24).

Indigenous peoples as minorities in settler states will always have to negotiate rights and interests *within* nation states and with the institutions of those states, but that does not mean that Indigenous peoples' claims can only be recognised *within* modern thinking. Modern knowledge, influential though it is, is neither omnipotent nor eternal. Indigenous people keep challenging the moderns with their cultural recognition claims, and, as Mexican poet Octavio Paz has argued, we should celebrate this diversity (Paz 1967):

> What sets worlds in motion is the interplay of differences, their attractions and repulsions. Life is plurality, death is uniformity. By suppressing differences and peculiarities, by eliminating different civilisations and cultures, progress weakens life and favors death. The ideal of a single civilization for everyone, implicit in the cult of progress and technique, impoverishes and mutilates us. Every view of the world that becomes extinct, every culture that disappears, diminishes a possibility of life.

The argument Octavio Paz makes about the diminishment of life and cultural uniformity is linked to Robert Pyle's argument about the extinction of experience and species loss. When an entire inland riverine ecosystem is under threat it is not just the species that are threatened, and the personal relationships with those species, but the cultural diversity that is intertwined with those ecologies.

From a viewpoint that emphasises connectivity, natural and cultural diversity sustain the plurality of life, and the life-sustaining relationships people hold with their local ecologies are now acknowledged and valued. This life is an important part of the negotiations that are held about water management. Natural and cultural diversity can form the meaningful base for an ecological dialogue that includes the diversity of lives and agencies in the inland river country.

INDIGENOUS PEOPLES' INVOLVEMENT IN MODERN WATER MANAGEMENT

Water is critical for Aboriginal people living in the Murray–Darling Basin, but they have only very recently been explicitly involved in government water management policies. Indeed, Indigenous peoples' political–legal water territorialities are only just starting to be acknowledged.

In Australia, centralised water management policy began in Victoria in the late nineteenth century when the colony passed legislation vesting it with the control of water. This was consolidated on a wider scale in 1915, when the states and Commonwealth signed the River Murray Agreement. These acts were only possible through the erasure of the political–legal territories of the Aboriginal people, and were repetitions of earlier erasures. For example, Indigenous people had been excluded from the self-governing polities that came together to form Australia as a federation of colonies; indeed, this federation was premised on their exclusion (Dodson & Strelein 2001; Brennan et al. 2005, p. 56). With repetition, the modern concepts of nation and nature have colluded to normalise this political–legal arrangement as pre-existing its representation. This is why native title was so controversial among mainstream Australia when it was belatedly recognised by the High Court in 1992.

When modern water policy-makers began to acknowledge the importance of 'community' values, Indigenous peoples' values were also included. For example, in the 1987 'Murray–Darling Basin Environmental Resources Study' Aboriginal peoples' water issues are illustrated by small summaries written by academics, and the retelling of the Ngarrindjeri creation story about the Murray River. The interests of Aboriginal people are included within the chapter titled 'Cultural heritage' (MDBMC 1987, p. iii and chapter 8). This chapter is focused on Aboriginal site identification and protection — such as middens and scar trees — and the importance of community involvement in site management. Aboriginal sites are important places that speak to the long history of Aboriginal peoples' occupation of the inland river country. However, this 'tangible objects' approach avoids engaging with complex and interconnected intangibles (English 2002, p. 219), such as relationships held in water and culture. The cultural heritage approach is incredibly limiting for the range of concerns Aboriginal people have about the health of the inland rivers.

At the same time as water policy broadened to involve community participation, Indigenous people made advances in their formal recognition of their governance responsibilities to country, including the joint-management of national parks from the mid-1970s (see, for example, Birckhead et al. 1993, Baker et al. 2001 and Howitt 2001). The recognition of native title has had a profound influence in raising the profile of traditional owners more broadly in government policy. In 2000, as part of the implementation of the *Environment Protection and Biodiversity Conservation Act 1999* (Cwlth), a National Indigenous Advisory Committee was established. Recognition of the rights and interests of Indigenous peoples' specific water issues have been slower, partly because water management is still perceived as a largely technical issue. A 2002 report on integrated catchment management reported that Indigenous peoples' involvement had been neglected by government, and that Indigenous peoples had been engaged with catchment management 'from outside the system' (Bellamy et al. 2002, p. 54). In 2001 the (former) Aboriginal and Torres Strait Islander Commission and the Lingiari Foundation established a water rights project and later published papers arguing for increased recognition of Indigenous peoples' water rights (Lingiari Foundation 2002; ATSIC 2002a and ATSIC 2002b). And, as mentioned earlier, in 2003 MLDRIN requested a discussion paper on water rights (Morgan et al. 2004).

In 2004 Indigenous peoples' water issues were formally recognised in the National Water Initiative (NWI) (COAG 2004; original emphasis):

52. The Parties will provide for indigenous access to water resources, in accordance with relevant Commonwealth, State and Territory legislation, through planning processes that ensure:
    i) inclusion of indigenous representation in water planning wherever possible; and
    ii) *water plans* will incorporate indigenous social, spiritual and customary objectives and strategies for achieving these objectives wherever they can be developed.
53. Water planning processes will take account of the possible existence of native title rights to water in the catchment or aquifer area. The Parties note that plans may need to allocate water to native title holders following the recognition of native title rights in water under the *Commonwealth Native Title Act 1993*.
54. Water allocated to native title holders for traditional cultural purposes will be accounted for.

This advice illustrates how policy has become more sophisticated than the early focus on 'site management' and 'cultural heritage'. Importantly, this high level of recognition critically shifts the pressure placed on Indigenous people away from arguing why they should be involved, to engaging with how that involvement can take place. This NWI advice is vital for informing water negotiation and planning where Indigenous people have not been invited to participate. The NWI directs that actions to include Indigenous water issues in water planning are to take place immediately and to be the responsibility of the states and territories (COAG 2004, Schedule A).

The policy advice also acknowledges water allocations for native title holders. The NWI is required to recognise native title interests in water by both common and statutory law. This may cast traditional owners who do not apply for native title, or are unsuccessful in their native title application, as illegitimate participants. (For a discussion on native title and the traditional owner identity see Weir & Ross 2007.) However, the NWI also places emphasis on Indigenous water issues outside of native title. This reflects the national trend for agreement-making based on building relationships between parties rather than relying on legal authorities (Bradfield 2004, p. 1). However, the term 'traditional cultural purposes' could limit the recognition of Indigenous peoples' water issues, according to interpretations of 'tradition' and 'cultural purposes'.

As cited above, the language of the NWI is also circumscribed by explicit qualifications about Indigenous representation 'wherever possible', and the incorporation of Indigenous objectives 'wherever they can be developed'. These are negotiations of power. The language suggests incapacity to include Indigenous representation and objectives in water management — either on the governments' side, the Indigenous peoples' side, or both. Geographer Sue Jackson and Wardaman and Torres Strait Islander man Joe Morrison critiqued the potential of the NWI to benefit Indigenous people, raising similar issues (Jackson & Morrison 2007, pp. 23–4). They are uneasy about the lack of engagement with Indigenous people on these matters, and how these policies will be implemented given the discretionary terms, the technical difficulties of water management, and the lack of capacity for this work among Indigenous people and government agencies (ibid., p. 24).

The states and territories have been slow to develop legislation that includes Indigenous peoples' water issues. Significantly, the only state water legislation in the Basin that recognises Aboriginal peoples' water issues also mentions their economic values (Productivity Commission 2003, p. 59). The objects of the *Water Management Act 2004* (NSW) include (Section 3; my emphasis):

(c) to recognise and foster the significant social and economic benefits to the State that result from the sustainable and efficient use of water, including:

...

(iii) benefits to culture and heritage, and

(iv) benefits to the Aboriginal people in relation to their spiritual, social, customary and *economic* use of land and water.

There is little legislative force in the Act to support these objects. The provisions relating specifically to Aboriginal people are limited to native title holders for their 'domestic and traditional purposes' (Section 55). To supplement the Act, the New South Wales Government established an Aboriginal water trust with $5 million to spend over two years on Aboriginal water-reliant enterprises (DAA 2004; see also Behrendt & Thompson 2003, pp. 71–2). The water trust supports market-based enterprises, and is blind to traditional authority structures.

With respect to the Murray–Darling Basin Initiative, engagement with Indigenous people has increased markedly in the twenty-first century. Indigenous people were part of The Living Murray consultation process, as discussed in Chapter 1 and later in Chapter 5. The Ministerial Council's policy on integrated catchment management specifically mentions Indigenous people as part of the community of stakeholders in the Basin (MDBMC 2001, p. 4). Indigenous people have become an acknowledged part of government water management, law and policy. Further, in 2004–05 a catchment management authority in New South Wales helped organise the first cultural water allocation in Australia. Under the Murrumbidgee Water Sharing Plan, water was directed to a wetland by the traditional owners' reference group that works with the Murrumbidgee Catchment Management Authority (DIPNR 2004, 3.3(c)).

Indigenous people have been instrumental in increasing this engagement with governments, particularly with the Murray–Darling Basin Initiative, and setting their own terms of engagement. The traditional owners have formed an alliance along the river country: the Murray Lower Darling Rivers Indigenous Nations (MLDRIN, see following chapter). They have also developed translation tools to negotiate with policy-makers, including the concept of 'cultural flows' (Morgan et al. 2004; Behrendt & Thompson 2003). Cultural flows build on the policy path made already by the advocates for environmental flows. The traditional owners have talked about the way cultural flows would return water to places of cultural significance, places where particular plants and medicines grow, or places where there are important Dreaming stories. The traditional owners are also talking about building a future for themselves based on their cultural connections with the rivers. As Yorta Yorta Elder Henry Atkinson said:

> As an Indigenous person, the environment, which includes the forests, the lands, the whole lot is the core of our very being. It is our past and our future.
> [7 August 2004]

One of their main arguments for cultural flows is that the water is critical for their cultural survival. The tradition owners, like Henry, express relationships with water through the language of connectivity. With the diminishment of those connectivities, their culture and identity as traditional owners is also diminished.

The traditional owners bring their language of connectivity to a negotiation table that has many challenges for them, but also opportunities. They must take on these challenges because of the political reality that the colonial power is unwilling to cede sovereignty over land and resources (Hunn et al. 2003, p. 80). The option of walking away from negotiations is always available, but in the Murray–Darling Basin the traditional owners choose to participate so as to transform the other actors. Despite obvious inequities in power, dialogue is an important part of influencing these actors and thus influencing water management responses to ecological devastation (see also Hunn et al. 2003, p. 80; Palmer 2006, pp. 34, 39).

The construction of Indigenous people as premoderns without history and western people as modern with history creates numerous difficulties for Indigenous people in these engagements. With an understanding of culture as process and practice, the dialogue between Indigenous peoples and governments can move beyond measurements of tradition to engaging with contemporary identities and practices. This is a different way of bringing the 'rights holder' into being. But this dialogue needs to expand further to include connectivity, if we want to really understand what the traditional owners are saying.

CHAPTER 5

Murray Lower Darling Rivers Indigenous Nations

Years ago they didn't give two hoots about what we said, but now they ask for our input.
                                                        Robert Charles, Wamba Wamba Elder

By the early twenty-first century, the federal government had revised water management policies to include the participation of Indigenous people and their water issues. But the traditional owners are not waiting for different layers of government to develop models for engagement; instead they have their own organisational approach. They have mobilised as an alliance in the southern part of the Murray–Darling Basin: the Murray Lower Darling Rivers Indigenous Nations (MLDRIN). MLDRIN is a specific intervention by the traditional owners in water law, policy and management. It is also a positive expression of their political identity.

The MLDRIN delegates are working to build the capacity of the nation groups, to be more self-determining and proactively, rather than reactively, engaged with other organisations. However, by getting so involved in policy the delegates are drawn deeper into transformative intercultural engagements. They have also developed new knowledge as cultural adaptations to this context. In this chapter, I choose not to focus on theoretical analysis but instead provide space for the delegates to describe their work in their own words.

CREATING A PLATFORM FOR ENGAGEMENT AND ACTIVITY

The formation of MLDRIN developed out of political action by the Yorta Yorta. In 1994 the Yorta Yorta people lodged their native title claim in the Federal Court. More than 400 parties opposed the claim, including the Victorian, New South Wales and South Australian governments, local governments, the National Farmers Federation, and the Murray–Darling Basin Commission (the Commission). After their initial loss in the Federal Court in October 1997, and in relation to a series of

---

1. Interviewed 24 June 2004.

other events, in August 1998 the Yorta Yorta people hosted a meeting of traditional owners from along the Murray River (see also Morgan et al. 2006, pp. 142–3; Weir & Ross 2007). Henry Atkinson, chair at that meeting, describes it thus:

> We had a bit of a discussion taking about two days to see what options were available for the Indigenous people to be able to have a voice within their traditional country in relation to the rivers and sacred sites and if possible arrange for Indigenous people to somehow get economic benefits for their natural resource … We spoke about different things like trying to protect the environment and for all the flora and fauna. We invited Mr Bob Smith; I think he was the chief executive officer of the Murray–Darling Basin at the time. He came along and fully supported what we were trying to achieve, and eventually the Murray–Darling Basin [Commission] decided to fund MLDRIN, even though during the native title trial of the Yorta Yorta case, the Murray–Darling Basin were really fighting against us to have say or ownership to some parts of the water. The Murray–Darling somehow or other changed their ways and supported all of the Indigenous groups that were there at the time sitting at the table.

This meeting provided the opportunity for a very different engagement between Indigenous people and government institutions. The parties had moved from adversarial litigation to sitting down together at a negotiation table. The Indigenous people drew on their self-determining frameworks and processes to call the meeting of traditional owners; the courts did not determine the 'traditional identities' present. Their authority to call this meeting was both recognised and supported by the Commission. Moreover, the engagement became an opportunity for dialogue which transformed both parties. Both parties emphasised their shared interests, were able to identify common ground and ways to work together. Native title was a transformative part of this process. The focus on positive possibilities by government and the traditional owners was in response to the limitations of native title, and also its existence.

A broader consultation process with traditional owners followed the first meeting in the Barmah forest. Yorta Yorta woman Monica Morgan (working for the Yorta Yorta Nation Aboriginal Corporation) and Mutti Mutti Elder Jeanette Crew (working for the New South Wales Department of Land and Water Conservation) undertook this work. With support from the traditional owners, the inaugural meeting of MLDRIN was held in July 2001 in Dareton, Barkandji country, close to the Darling's junction with the Murray. Jeanette Crew spoke to me about the empowering buzz she felt at the meeting in Dareton:

5. MURRAY LOWER DARLING RIVERS INDIGENOUS NATIONS

… historically it was probably the first time, not just in historical times since whitefella came, but ever, that all people from all those nations were together in one room at the same time.

MLDRIN's current executive officer Steven Ross, who is Jeanette's son, told me about his experience attending the Dareton meeting while working for the NSW Department of Aboriginal Affairs:

… people talking about quite technical matters in an Indigenous way and making some important decisions about what they wanted to do with the rivers and the connected lands, the surrounding environment. It was really powerful.

Since then, the traditional owners have consolidated their organisational structure and election processes, and have incorporated under the *Corporations Act 2001* (Cwlth). The traditional owners describe MLDRIN as a 'confederation' of traditional owners or Indigenous Nations from the river country. They meet about four times a year as a board of delegates, with two delegates for each nation. A working group of five people meets more often.

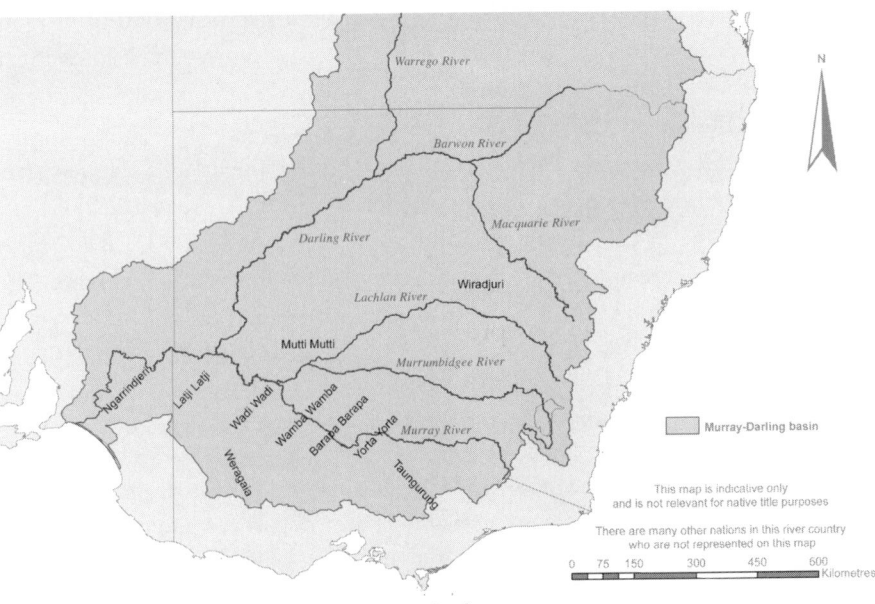

The Indigenous Nations who have formed the alliance with the Murray Lower Darling Rivers Indigenous Nations.

There are currently 10 nations in the confederation: the Wiradjuri, Yorta Yorta, Taungurung, Wamba Wamba, Barapa Barapa, Mutti Mutti, Wadi Wadi, Latji Latji, Wergaia and Ngarrindjeri nations (see following figure.). Their country runs from the headwaters of the Murray in the eastern mountains, and the Macquarie River in the central north, and then along the length of the rivers to the Murray Mouth in South Australia.

These nations are mostly, but not exclusively, neighbours along the Murray. The country of the traditional owners can be on either side of the rivers, as well as crossing state boundaries. Country boundaries are rarely closely defined. There is also shared country, whereby two or more nations have authority to speak for the same country.

Crucial to the confederation's ongoing operation, the MLDRIN delegates have been able to secure its funding base. In 2003, the MLDRIN delegates entered into a three-year funding agreement with the Murray–Darling Basin Commission which included funding for meetings (travel, accommodation, meals and meeting room costs) and a full-time executive officer. This arrangement was renewed in 2006 for another three years. MLDRIN delegates do not receive sitting fees, which for some delegates is a matter of principle, to reduce the risk of co-option of their independence. However, the alliance has negotiated a service delivery contract with the Murray–Darling Basin Commission for when they give knowledge and advice about The Living Murray.

As part of becoming an incorporated legal entity the MLDRIN delegates were required to prepare a constitution, which includes an explanation of their objects. According to the constitution, the main object of MLDRIN is to act as trustee to two separate funds: the Cultural Fund and the Environmental Fund. The corporations legislation made it difficult to establish a fund that was both cultural and environmental, and MLDRIN had legal advice recommending against trying to create one fund. The purpose of these funds is to (Constitution of Murray Lower Darling Rivers Indigenous Nations 2007, *Corporations Act 2001* (Cwlth), p. 5):

(a) confederate the Indigenous Nations within the Traditional Region, each with an inherent and unique connection to their respective traditional lands and waters, so that the Indigenous Nations can better work together for the protection and enhancement of the natural and cultural environment of each of their peoples and their territorial lands and waters;

(b) create a forum for the Indigenous Nations to speak about their peoples' inherent environmental and cultural heritage rights, in respect of natural and cultural resources, with a focus on water, by ensuring their real participation in resources management;

(c) establish mechanisms to sustain the Indigenous Nations' interests by which the physical, spiritual, cultural, social and economic interests in the lands and waters of the Traditional Region that will ensure the continuation of the peoples of each Indigenous Nation and their ability to enjoy the natural and cultural resources of the Traditional Region, an ability which is conditioned upon the integrity and aliveness of the biodiversity of flower, fauna and aquatic life;

(d) raise funds to further the objects of the Environmental Fund and the Cultural Fund; and

(e) do all other things and exercise all the powers set out in the Cultural Fund Trust Deed and the Environmental Fund Trust Deed and this Constitution that are necessary to reach the objects of the Company, the Cultural Fund and the Environmental Fund.

The delegates have also established <www.mldrin.org.au> as MLDRIN's web presence. This website displays their logo, the mussel. The text on the website describes this choice (MLDRIN 2008):

> The mussel was selected because it is a creature that is often overlooked in the rivers, despite it being an important indicator of the health of the river. The mussel also holds special cultural meaning for many Indigenous Nations — as Aunty Mary Pappin of the Mutti Mutti stated:
> 'Everyone gets excited when they come across an old midden site and the one thing people always find in middens is mussel shells. It was always an important food source for our people.'

The energy and direction of the traditional owners involved with MLDRIN shows how they are strongly motivated to restore river health. They are doing so in very modern and Aboriginal ways. The differences in language between the constitution rules and their website show how the delegates adopt a more formal language for legal documents. Adopting different languages is part of engaging with different institutions and media, and is a strategic choice.

NEGOTIATIONS AND POLITICAL IDENTITY

The creation of MLDRIN by the traditional owners is based in their political identity. As First Nations, Indigenous people in Australia have systems of law and governance in connection to country that existed before colonisation. Despite this, modern knowledge presumes a homogenous national citizenship, thus denying the political–legal territorialities of Indigenous people in settler states. Within modern approaches to governance, Indigenous people are part of the citizenship only as a special interest group. Indeed, in Australia prior to 1967 Indigenous people were not even included as citizens. The discriminatory theory of hierarchical civilisation, whereby Indigenous people were at a lower stage of civilization than the moderns, meant that Indigenous people were treated as wards of the state. The creation of MLDRIN by the traditional owners is a deliberate counter to the assumed political uniformity of national citizenship.

The delegates emphasise MLDRIN as an alliance of political entities by their choice of the word 'confederation'. The delegates further stress that MLDRIN is not a substitute for the authority of the traditional owners. MLDRIN is an umbrella organisation designed to create a greater voice for these nations, and to help them build their own capacity to engage with governments at all levels. The authority rests with the traditional owner groups. Thus, the effective operation of MLDRIN is reliant on good communication with each Indigenous nation through their delegates, and good governance skills within the nations so that the delegates are accountable to their nation (see further Weir & Ross 2007). In line with this, the MLDRIN delegates negotiated a governance project in 2007 with the Commission to provide basic funding for the traditional owners to organise meetings at a nation level.

The MLDRIN delegates are also involved in succession planning to pass on their responsibilities to the next generation. They successfully applied for Commonwealth funding to train young traditional owners in their responsibilities to country and to their nation group, running the first 'Cultural Connections' course in 2007. Delegates also mentor younger family members by sometimes bringing them to meetings as observers.

The formation and work of MLDRIN is what Anna Tsing would call a creative and self-conscious scale-making project, both practised and produced (Tsing 2005, pp. 58–60). The scale of relationships between traditional owners and country is important, but not large enough to engage with the large Murray–Darling Basin administration area. Through scale-making, the traditional owners have built a regional, state and national dialogue while continuing to emphasise their specific knowledge of country. The delegates carry this specificity through the scale-making project, articulating their issues so that they become collapsible and compatible across different scales. This increases the potential for their arguments to be adopted into government policy.

Yorta Yorta woman Monica Morgan spoke to me about the connections within the alliance:

> You've got friendships, you've got families, you've got trade — all these things that have been there for thousands of years and through the colonisation process of course we have helped each other. When we're on the run from Yorta Yorta country, go up to Wamba Wamba — you know ... we share – there is no fence that goes up that stops the water flowing or things going all around, we affect each other. [1 July 2004]

Ngarrindjeri Elder Agnes Rigney evokes the rivers to describe the connections held between the nations:

> ... it doesn't matter what language group or what group you belong to, there is a common cause, there is a common thread along the river, the river people. [21 July 2004]

The MLDRIN delegates' commitment to building a solid alliance across the length of the river country also reduces the grounds for others to intervene in MLDRIN business with respect to questions of traditional identity. By negotiating these matters internally with each other, the MLDRIN alliance relieves the traditional owners *and* government from public determinations of traditional identities. However, this does not mean that the complex identity issues no longer exist (the intra-Aboriginal politics inherent in the formation and operation of MLDRIN is discussed in a paper co-authored with MLDRIN executive officer Steven Ross (Weir & Ross 2007)).

Some of the people involved in MLDRIN have a vision of all the Indigenous nations in the Basin coming together to negotiate with government on water policy and other matters. In May 2004, when Monica Morgan was working for the Commission, the Commission funded the first ever 'Indigenous Basin-Wide Gathering' in Canberra (MDBC n.d., p. 3). There are about 40 traditional owner groups in the Basin, about 27 of which attended the gathering. At this meeting, the participants proposed that a northern alliance form to represent traditional owners from the northern part of the Basin, and that MLDRIN work to include all the nations from the southern part. To facilitate this, a working group took a red marker to a Commission map of the Basin and divided it into northern and southern sections. However, the process was to prove slower and more complex in reality. The traditional owners from the north have not mobilised themselves, whilst the MLDRIN alliance has focused on consolidating the capacity of its membership of 10 nations. The alliance of 10 nations is not a permanent arrangement, and other nations have been active at MLDRIN meetings. For example, the Barkandji and Nyampaa nations (New South Wales) were more involved in the early period

MLDRIN delegates and observers at a MLDRIN meeting, July 2004, Deniliquin, Wamba Wamba / Barapa Barapa shared country.

of MLDRIN activities but their involvement declined. At the 2004 Gathering, Barkandji advised that they would form a separate relationship directly with water management bureaucracies rather than have their relationship 'mediated' through MLDRIN.

The importance of developing networks of traditional owners and mobilising their activity around a range of concerns is something Henry Atkinson spoke to me about. He recalled what he said to the first meeting in the Barmah forest:

> I said I imagined throwing a stone in a pool of water, and with the ripples going out, those ripples getting bigger and bigger, which means that MLDRIN would be getting bigger, bringing those people together for the benefit of all peoples. And I imagined MLDRIN being a stepping stone, starting off with the water issue and then being able to be a stronger voice, not just for the water, but for other issues that will evolve into the future … I just hope that they realise that the stronger they get, the bigger the voice they have. [7 August 2004]

Mutti Mutti Elder Mary Pappin, Wamba Wamba woman Tracy Hamilton, Mutti Mutti Elder Jeanette Crew, Wamba Wamba man Steven Ross, Yorta Yorta Elder Henry Atkinson, October 2004, Macquarie River, Dubbo, Wiradjuri country.

The creation of MLDRIN was also a challenge by the traditional owners to the portrayal of Indigenous identities as homogenous. Aboriginal people with diverse identities live together within the contemporary intercultural communities now established on country. The 'historical people' hold important historical ties where they live. The 'stolen generations' are people who were taken from their families as children, and who now may or may not know where their traditional country is. There are also people who live physically apart from their traditional country but maintain a strong sense of connection and knowledge of their country (Smith 2000, p. 3). As part of these diverse Indigenous identities, MLDRIN delegates are keen to establish their identity as traditional owners, whilst also acknowledging the common issues faced by all Indigenous peoples. The delegates identify country as specifically the business of the nations, as distinct to citizenship issues facing Indigenous people more generally, such as health, housing and education (Morgan et al. 2004, p. 22). Being politically astute, in their documents the MLDRIN delegates declare that the agreements and relationships they make with government are non-exclusive (see, for example, Clause 3 of their memorandum of understanding with the Commission). That is, other Indigenous organisations and groups can build relationships with government independent to MLDRIN. However, at the same time the delegates assert that only traditional owners can speak for country.

Despite assertions of the independent authority and identity of the nation groups within the MLDRIN alliance, the priorities of MLDRIN are influencing the business of the traditional owner groups. This influence could potentially add to tensions about the authority of MLDRIN, given the expressed importance by the traditional owners of localism as the source of governance and authority (see also Smith 2005).

The MLDRIN delegates and the natural resource managers and policy-makers they work with are also transforming each other. The delegates' arguments about political identity are changing the way Indigenous consultation and negotiation in natural resource management is conducted by government. Likewise, the close engagement the traditional owners have with government is transforming the way MLDRIN and the nations manage their own concerns.

EXTENDING THE PRINCIPLES

In establishing the alliance, the MLDRIN delegates sought to extend the principle that traditional owners have a distinct political status, ongoing since before colonisation, and thus are not just one among many stakeholders in the model of natural resource management consultation. They do this by developing distinct partnerships with a range of governments, research institutions and environmental non-government organisations. These partnerships all contain acknowledgments of the traditional owners, their specific relationships with country, and the importance of their decision-making structures. By prioritising 'process', the delegates build relationships that then have the capacity to address the complex issues that are brought to the partnerships.

In the early written material by MLDRIN the authors criticised past natural resource management consultation with Indigenous people as 'non-existent, half-hearted or inappropriate' (MLDRIN n.d., p. 1). Sometimes government consultation on country matters occurs with Indigenous people who are not traditional owners of country, even on issues as central to traditional owner identity as cultural heri-tage and sites of significance. There are also unrealistic expectations by natural resource managers that one Indigenous representative on a natural resource management board is adequate representation. Instead, the traditional owners argue that Indigenous engagement should occur directly with the governance structures of the traditional owner groups. There are a plethora of governance structures in natural resource management, and the MLDRIN delegates work to counter-assert their governance structures in each context.

The influence of these arguments, concomitant with the profound legal and philosophical changes wrought by the recognition of native title in *Mabo*, is now evident in relation to natural resource management for nation groups in the Basin. The North Central Catchment Management Authority in Victoria has signed a protocol agreement with the North West Nations and the Yorta Yorta Nation (NCCMA 2002). The Murrumbidgee Catchment Management Authority in New South Wales has many Indigenous nations within its catchment boundaries, and has supported the establishment of a 'Traditional Natural Resource and Cultural Heritage Reference Group' comprised of Mutti Mutti, Nari Nari, Wiradjuri and Ngunawal Elders (CMA 2004, p. 42). The MLDRIN delegates pursue similar partnerships that emphasise process, albeit on a larger scale. When the traditional owners were setting up MLDRIN they proposed an agreement between MLDRIN and the New South Wales Department of Land and Water Conservation. Subsequently, a Memorandum of Understanding (MoU) was signed between the two parties in October 2001. After four years of negotiation with the Murray–Darling Basin Commissioners, MLDRIN delegates also signed an MoU with the Commission in 2006. Both MoUs speak in general terms and identify issues on which the parties will cooperate on together. These include, but are not limited to, representation and participation, cultural heritage, and natural resource management. Neither MoU is legally binding, nor do they recognise or grant legal rights in land or water for the traditional owners; nevertheless, these memoranda include substantial achievements about process. The parties recognise each other, agree to develop frameworks and processes to reach understandings and agreements on issues of common concern, and will work to involve the traditional owners in natural resource management of the Murray–Darling Basin.

The MLDRIN delegates are also working to build partnerships with other organisations, such as the Commonwealth Scientific and Industrial Research Organisation (CSIRO) and environmental groups. CSIRO has undertaken research into 'Indigenous values of water' with the Ngarrindjeri nation, and may undertake research into the effects of climate change on traditional owners in the Basin, both research requests from the MLDRIN delegates. The MLDRIN delegates also signed a cooperative agreement in 2007 with representatives of several environmental non-government organisations: the National Parks Association of NSW, Friends of the Earth, The Wilderness Society, the Victorian National Parks Association, the Australian Conservation Foundation, Environment Victoria, and the Nature

Ian Sinclair, President of the Murray–Darling Basin Commission and Ngarrindjeri Elder Matt Rigney, Chair of MLDRIN, hold up the signed Memorandum of Understanding between MLDRIN and the Murray–Darling Basin Commission, 24 March 2006, Wonga Wetlands near Albury, New South Wales, Wiradjuri country.

Conservation Council of NSW. This agreement recognised the Indigenous nations as the traditional owners of country and their inherent right to speak for country. The agreement also recognised the role of environmental groups in representing the concerns of their membership, and the role they play in the creation of an equitable, healthy and sustainable country.

Thus, the traditional owners prioritise working out the terms of the negotiation table up-front, to establish themselves as polities in their own right. However, entering into such agreements is a transformative process as each party strives to influence the other in order to reach common ground. Compromises are always a part of this. Formalised in the document, the shared ground becomes meaningful to both parties. The document becomes a reference point for the relationship

Wiradjuri Elders Ramsay Freeman and Tony Peachey at the signing of the Memorandum of Understanding.

and the parties' responsibilities to each other. The compromises involved in the negotiations means that they are not final perfect agreements, and should always be open to review as part of accepting the dynamism of the partnership (Tully 2004b, p. 98).

The MoU between MLDRIN and the Commission is a negotiated outcome in which both parties had points they were willing and not willing to compromise on. For example, the Commissioners were unable to sign off on a document which spelt nation with a capital 'N'. The MLDRIN delegates and the Commissioners eventually agreed upon the compromise 'Indigenous nations'. Much later on in the negotiations a newly appointed Commissioner requested that 'Indigenous nations' be changed to 'Indigenous peoples'. However, MLDRIN was not prepared to compromise on this.

The disagreement over n/Nation and Indigenous peoples/nations centres on the political identity of the traditional owner groups (and note that not all of the traditional owner groups within the alliance use the term 'Nation' to describe themselves). The term 'nation' as a description of the social and cultural organisation of peoples comes from the early anthropologists who used nation for Aboriginal cultural blocs formed on the basis of genealogical descent. At that time nation did not carry the implications about sovereignty that have developed since the early twentieth century (Blackburn 2002, pp. 150, 153). Today, by describing themselves as Nations (emphasised by the capital letter), the MLDRIN delegates assert the sovereign references now linked to that term. This includes the assertion of political autonomy, internal self-government, and external self-determination (Strelein 2002–03). They are also connecting their experiences to the Indigenous rights struggles of First Nations peoples in North America.

The different languages and values embedded in the terms 'Indigenous Nations', 'Indigenous nations', and 'Indigenous peoples' were critical to each party. For the traditional owners, it was about their identity and their ability to define themselves. I am assuming that the Commissioners, who are high-level public servants, were concerned about the modern universal understandings of 'nation'. Such understandings grasp the assertions of sovereignty by the traditional owners as a threat to the national identity or legal identity. Indeed, the Commissioners requested the inclusion of a clause that tells us that the memorandum 'is not a treaty' (Clause 9.1.6.).

The language of the MoU shows evidence of how both parties work to make it meaningful to them. Consider Clause 5.1, the 'Nature of Relationship':

> The Parties agree that by this Memorandum of Understanding they establish a cooperative relationship, to help ensure that the natural resources of the Murray and Darling River valleys are managed in a manner which recognises and ensures that the 'efficient and sustainable' use of those resources carries benefits to the Indigenous nations in relation to their cultural heritage, including without limitation, their spiritual, social, customary and economic values for the use of land and water.

This paragraph manages to communicate matters of importance to both parties. It speaks of natural resources in terms of use and management, including 'efficient and sustainable' as a clear reference to the importance of the ongoing productivity of rural industries. This clause also relates that this productivity should carry benefits — without limitations — to the holistic values of the Indigenous nations. Thus, it links modern thinking and Indigenous peoples' holistic knowledge, revealing the complexity of intercultural engagements, and transformations in both Indigenous and modern knowledge traditions.

As agreements can never be perfect or final, disagreement will always continue, but what is important is the process: that both parties are satisfied they have been treated with respect. This shared experience is an investment in a partnership that can then address challenging issues when they arise (Tully 2004b, p. 100). Moreover, encounters of difference within a common cause are fertile collaborations. The common cause is the cultural encounter, in which new meanings and ideas are productively discussed and developed. Indeed, Anna Tsing has argued that agreement is 'most successful' when this vision appeals 'simultaneously to divergent cultural legacies' (Tsing 2005, pp. 246–7). Finding agreement across cultural difference strengthens commitment to the common cause of the collaboration. Commitment to a common cause is evident in the Murray–Darling Basin Initiative's The Living Murray.

IN BED WITH THE LIVING MURRAY

The Ministerial Council established The Living Murray initiative in mid-2002 to return water to the river as 'environmental water'. It is a major undertaking, with $1 billion committed at the time by state, federal and territory governments. MLDRIN supported an 'Indigenous Partnerships Program' to complement the work of The Living Murray, and to pursue their longer-term objectives for the nations and country.

During 2002 and 2003, Commission staff conducted extensive community consultations to engage with the concerns of water entitlement holders about whether returning water to the rivers would have negative consequences for their water allocations. The community consultation process for The Living Murray discussed three reference points for the return of water: 350 gigalitres, 750 gigalitres or 1500 gigalitres (MDBMC 2002, pp. 2–3). The MDBC staff (partly) allayed the concerns of water entitlement holders by emphasising that this 'First Step' would source extra water through (MDBMC 2003a, p. 2):

> ... on-farm initiatives, efficiency gains, infrastructure improvements and rationalisation, and market-based approaches, and purchase of water from willing sellers, rather than by way of compulsory acquisition.

As part of the consultation, Commission staff contracted a parallel process specifically with Indigenous people. This parallel consultation process reflected the learning the Commission had already achieved from their engagements with Indigenous people in the Basin (including the commissioned report by Forward

NRM & Arrilla-Aboriginal Training and Development 2003). Five central themes were identified in the Indigenous response to The Living Murray (FCG 2003, p. 4), which have been paraphrased as (Morgan et al. 2004, pp. 14–15):

- **A shared vision**: While recognising the diversity of views among Indigenous Nations and local communities, as well as the different perspectives from various state governments and non-indigenous communities, Indigenous peoples of the Murray seek a shared and integrated vision for a healthy river;
- **Recognition**: The report seeks recognition of the status of Indigenous Nations as peoples, and of their inherent rights to exercise their culture and sustain their communities on their traditional lands;
- **Respect for country**: The environmental health of the Murray River system is prioritised in the Report as it is integral to the cultural, social and economic health of Indigenous communities;
- **Involvement**: Throughout the report Indigenous peoples emphasised their desire to be actively involved at all levels of management of water and other natural resources on their traditional lands.
- **Policy change**: Indigenous peoples also proposed specific changes to policies central to The Living Murray initiative as well as a general change in approach toward a cultural and natural resource model.

The five themes show how the Indigenous people commit to common ground with the Commission, seek involvement in government programs, but also prioritise a change in the approach towards a 'cultural and natural resource model' (Morgan et al. 2004, p. 15).

A consistent response given in these consultations was that the reference figures for returning water to the environment that featured in the consultations were too low: scientific advice had recommended 4000 gigalitres for a 'high' chance of restoring river health and 1900 gigalitres for a 'moderate' chance (FCG 2003, p. 24; Jones et al. 2002, p. 27). After the consultations, the Ministerial Council decided that The Living Murray initiative would return 500 gigalitres as environmental flows by 2009. That goal continues, but took a new direction in November 2003 when the 'icon sites', or Significant Ecological Assets, were announced (MDBMC 2003a; MDBC 2005). The success of The Living Murray is measured by the environmental benefits achieved at six Significant Ecological Assets: the Barmah–Millewa forest; Gunbower and Koondrook–Perricoota forests; Hattah Lakes; Chowilla Floodplain (including Lindsay–Wallpolla); the Murray Mouth, Coorong and Lower Lakes; and the River Murray channel (see figure opposite). At these six icon sites, water flows are to be recovered and managed to increase the environmental rehabilitation of birdlife, vegetation, wetlands, aquatic vegetation and native fish, including

5. MURRAY LOWER DARLING RIVERS INDIGENOUS NATIONS

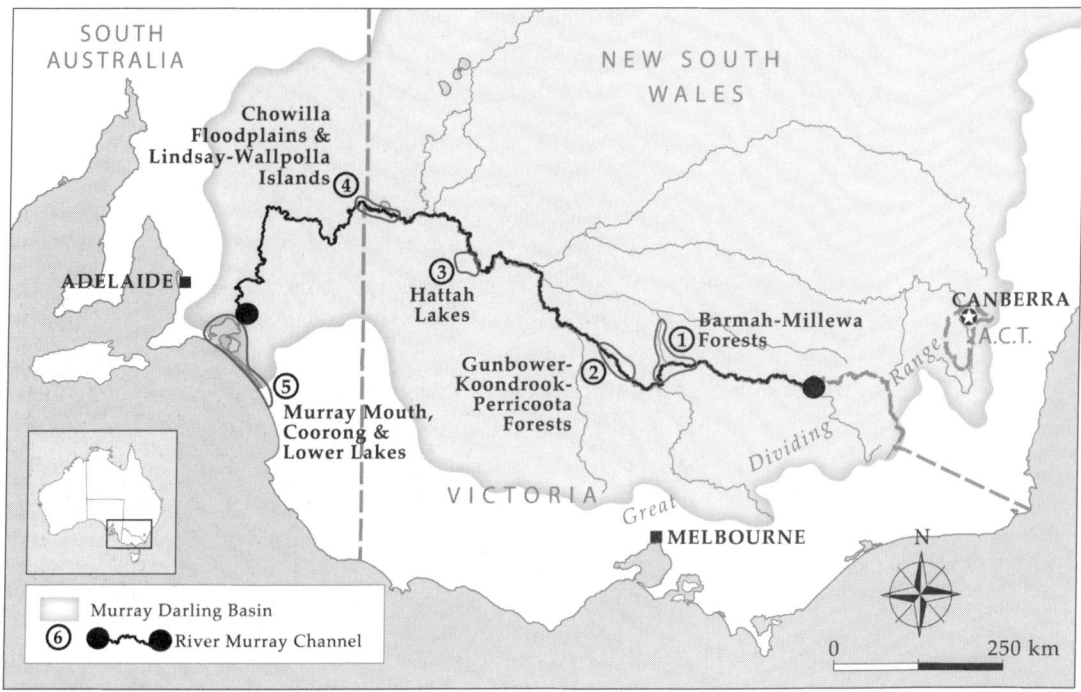

The six Significant Ecological Assets identified and managed as part of The Living Murray program (based on map © MDBMC 2003a).

maintaining the Murray River mouth (COAG 2004). An Indigenous Reference Group has been established at each icon site, with representation from traditional owners.

The Living Murray is the main environmental program being implemented by the Commission, and the MLDRIN delegates have committed a lot of energy to ensuring that they are part of this work. With the participating nations' support they developed the Indigenous Partnerships Project, which then became part of the Ministerial Council's First Step decision. This project is designed by the delegates to make The Living Murray responsive to the governance structures of the Indigenous nations (see further Morgan et al. 2006). The partnership project specifically targets the management plans of the icon sites, building in employment and resources for traditional owners at each place, including the development of maps that express activity that is meaningful to the traditional owners. The activities of individuals in their country, such as fishing and camping sites, are mapped and a compilation of those activities generates a larger-scale image of connections with country across a community. Canadian researcher Terry Tobias, who has experience creating 'land use and occupancy maps', was contracted to work with the nations (Tobias 2000). The MLDRIN delegates anticipate that the creation of such maps for each icon site will help communicate their relationships with country to government, and be an exercise in strengthening community identity and documenting local knowledge. A pilot mapping project with Terry Tobias was undertaken in the Barmah–Millewa icon site in Yorta Yorta country (note that Yorta Yorta country is much larger in area than the map produced for this icon site) (see also Ward 2009). Ideally, each nation would have the choice of getting involved in such a mapping exercise, and for the breadth of their country; however, the focus for now is on the icon sites, in accordance with the priorities of the Commission.

Land use and occupancy maps are simple but effective tools. The maps usefully focus on what people actually do, rather than trying to measure the authenticity of Indigenous peoples' relationships with country. They map direct connections to country, and they are a deliberate shift away from government policies that understand country through 'cultural heritage' and 'site management'. Lines on maps are just one way of representing the world, but this method is translated by the moderns into objective truths (Mitchell 2000, p. 15; Braun 2002, p. 43; Scott 1998, p. 49). Bruce Braun illustrated this with the way the 'culturally modified trees' map changed the conflict over the temperate rainforest in British Columbia (Braun 2002, p. 102). The traditional owners use modern map-making techniques to negotiate with modern institutions. In the longer term, the documents will carry historical value for the nations. The maps will inform Indigenous identity through the reflexive identity constructions of future generations of traditional owners.

Whilst the maps are popular, and the Indigenous Partnerships Project is a constructive step, there remain a series of challenges for MLDRIN delegates with The Living Murray, including the exclusive focus on the icon sites. As Matt Rigney has said (quoted in Weir 2006, p. 35):

> Those areas that aren't part of the top six aren't seen as important, it's only those sites of scientific importance that matter to the government.

Some assets are wholly within the country of one Indigenous nation, while others are on shared country or straddle the country of several nations. The MLDRIN delegates workshop the ongoing governance implications of this at their meetings and with their nation groups.

The delegates also have problems with the fundamental framework of The Living Murray. They perceive the goal of returning 500 gigalitres of water to the Murray by 2009 as grossly inadequate, and that the focus on infrastructure and water trading to recover water continues to divert attention away from directly addressing the over-allocation of water entitlements.

For the delegates, their work with The Living Murray has highlighted the different capacities of the two parties, and the reality of their negotiated relationship. For example, in a MLDRIN meeting in Echuca in 2004, a representative from the Commission gave a one-hour presentation to 'consult' with the delegates about The Living Murray 'environmental works and measures program' designed to improve environmental health at the icon sites (MDBC 2007i). In this hour, the representative ran through $150 million worth of spending on 38 different infrastructure projects, some of which had already started. The list of projects, littered with the jargon of natural resource management, was distributed among the participants. Both audience and presenter recognised that feedback was going to be totally unrealistic in this context. It is not enough for dialogue to exist, but for the dialogue to be cognisant and responsive to the styles and resources of both parties.

The scale of The Living Murray has made large demands on the work of the MLDRIN delegates, and since their intense engagement with The Living Murray in 2004, the MLDRIN executive officer has worked to ensure other priorities are not neglected. This work has included developing a strategic plan in 2006, holding workshops to mentor younger generations, lobbying for funding to support governance at the nation level, applying for grant monies to undertake activities beyond the priorities of the Commission, and hosting the *Dhungala Festival: The Spirit of Water* in 2008 to celebrate the tenth anniversary of MLDRIN.

TRANSLATION AND TRANSFORMATION

All engagements carry the potential for both positive and negative outcomes, as we are always transformed through our dialogues with each other. For the MLDRIN delegates, the differential in power dynamics in their engagements with government makes this a risky context. The less-powerful party carries a greater risk of being co-opted by the other party. Indeed, the close relationship that MLDRIN holds with the Murray–Darling Basin Commission has exposed it to criticisms of being an extension of government policy implementation.

To facilitate their vision for the restoration of the inland rivers, the MLDRIN delegates must necessarily engage with a myriad of other organisations concerned about river health. The MLDRIN delegates negotiate partnerships with these organisations, they populate positions on many water management boards and councils (including the Ministerial Council's Community Advisory Committee), and they have taken contract and paid employment with water bureaucracies. For example, in a deliberate exercise to exert influence from within, one of the founders of MLDRIN, Monica Morgan, applied for and accepted a position with the Commission for 18 months from 2003 to 2004, taking responsibility for community consultation processes for The Living Murray. Yet many of the organisations the MLDRIN delegates are engaging with have greater human and financial resources, as well as rights and responsibilities in relation to water that are entrenched in Australian law. Often the delegates feel swamped by the agendas of all these other actors. When the MLDRIN delegates translate their vision of restoring river health to these actors, the risk is that their message will not be heard, or will be transformed or possibly corrupted by dominant modern frameworks to such an extent as to render the message meaningless.

Consider the MoU between MLDRIN and the Commission. In Clause 4.1.4 both parties agreed to (my emphasis) 'ensure that the traditions of the Indigenous nations are *incorporated* into natural resource management policy development and implementation'. This contradicts what we know about connectivity: how the traditional owners carry knowledge of a distinctly different philosophical tradition that is central to their message for river restoration. These traditions have a very different conceptual framework to that of natural resource management. Yet here the traditional owners are actively suggesting that their traditions be incorporated into modern knowledge. This can be partly explained by the distinction between knowledge and practice. For example, I am told that the MLDRIN delegates speak about their identity as traditional custodians in their own space, but use the term 'traditional owners' in their engagements with government (Steven Ross, pers. comm., September 2007). Thus, the language of Clause 4.1.4 could be explained as a strategic compromise. Perhaps if natural resource management is also transforming as a result of engagements with Indigenous people and country, and in response to ecological devastation, then this could be a different natural resource management that the traditional owners are subscribing to.

Another example is native title. Since *Mabo,* the unique relationship between traditional owners and their country has gained wider currency in government policies. However, relationships between traditional owners and their country are not discretely definable interests, and yet this is what governments seek from native title. The *Native Title Act 1993* (Cwlth) requires traditional owners to provide a clear explanation of how the group is defined, what laws and customs unite them and the extent of their territory. Some state governments will not enter into mediation with claimants until they are convinced that all overlaps and disputes with other native title claimant groups are resolved. Governments extend this preference for certainty in relation to identity to certainty in relation to organisational form (see Smith 2003; Weir & Ross 2007). Native title holders are required to establish a corporation to manage transactions with other parties in relation to their native title rights and interests (Weir 2007a).

Across Australia, the formalisation and incorporation of organisational structures are now critical to how Indigenous peoples' rights are being recognised by governments (Rowse 2002, p. 179). This emphasis on certainty and identity has encouraged the Murray River traditional owners to create an organisation that offers just that. The MLDRIN delegates assert fixed relationships between traditional owners and their country, and have formalised these relationships in documents, public statements and agreements. In these communications the traditional owners describe a match between people–language–country, amongst a mosaic of countries, although they also promote an understanding of shared country. Whilst the MLDRIN delegates reinforce the trend of describing relationships between traditional owners and their country in a more bounded way, they keep the debate within the authority of Indigenous people and not subject to tests by external institutions that are unable to fully comprehend the systems they seek to 'assess' (Strelein 2000). In reality, relationships with country are complex, indeterminate, and multi-layered (Sutton 1995, pp. 49–50; see also Keen 2006, pp. 134–5). Shifts in identity are always occurring because relationships between people and their country are dynamic. They are influenced by many matters, including marriages, political manoeuvring, personal experiences, and the location of towns, mission lands and roads. Understandings of identity are also complicated by scale. Where does a family group, clan, tribe, nation or language group begin and end? (See further in Weir & Ross 2007.)

Both traditional owners and governments draw on modern assumptions about Indigenous people 'without history' to fix traditional identity as timeless — something that is inherited, and thus a non-negotiable fact. When shifts in identity occur, as they will from time to time, this fixed framework does not allow either party to interpret these shifts as a legitimate part of identity formation. The different alliances made over native title are a good example of contemporary influences that negotiate Indigenous identity. It is difficult to predict whether the future success or failure of lodged native title claims will cause readjustments and realignments in the MLDRIN alliance. However, the strategic compromises that the MLDRIN delegates make and adopt in their partnerships with government will reflexively influence the knowledge the practice stems from.

Because the delegates are pursuing so much from governments they are drawn into deeper intercultural engagements. Arguably, packaging their identity in a form preferred by government will increase the possibilities for the recognition of their rights. For example, as a legal entity it is now much more practical for MLDRIN to lobby for a water allocation. This certainty must be managed with respect to the diverse and dynamic identities of the nation groups, otherwise the MLDRIN delegates risk the filtering of their perspective and authority as traditional owners.

SPEAKING FOR COUNTRY

'Speaking for country' is an authority that the MLDRIN delegates assert as traditional owners. Steve Kinnane has written about the protocols of speaking for country. Steve Kinnane is a descendent of the stolen generations: his grandmother was taken as a child from her country in the Kimberley region of Western Australia. Today this country lies beneath the water storage known as Lake Argyle. He has written about this country (Kinnane 2002, p. 21):

> This is Miriuwung Country. I am linked through my grandmother to this place in many personal and tangible ways but I don't speak for this country. Speaking for country is not a simple thing. Only Law people (traditional owners) who have knowledge, specific relationships, and special rights and responsibilities can speak for country.

Thus, certain people who have knowledge and relationships with country, and with each other, can speak for country. These are intergenerational responsibilities. However, while a descendent of the genealogical group, Steve Kinnane has said that he does not have this authority. This authority is more than a birthright, it is something that must be known and practised. He calls traditional owners 'Law people', referring to the systems of laws and governance that are taught to certain people entrusted with speaking for country.

Mary Pappin talked to me about how these responsibilities are negotiated between neighbouring traditional owner groups. She spoke about the importance of lobbying government for water to flood creeks in Mutti Mutti country where bush foods and medicines used to grow. Mary was concerned that environmental flows might not come to Mutti Mutti country. In that eventuality, Mary discussed the political implications:

> ... we're going to have to go into areas where the environmental flows are to resource the country if the need is there. It might be going into somebody else's country, and I don't normally like that. But there are ways around that. To take something out of there, if you spoke to the Elders they would say 'Yes, you take what you want and then go'. [22 July 2004]

The one-time water pumping station at Raukkan, the former Point McLeay mission, on the banks of Lake Alexandrina, South Australia, Ngarrindjeri country.

The Pink Lake, South Australia where, a battle was held between the Ngarrindjeri people and the Ngarkat people — the pink is the blood of the warriors who were killed in the battle.

The Murrumbidgee River during drought and a dust storm, near Yass, New South Wales, Wiradjuri country.

The former wetland Psyche Bend Lagoon, near Mildura, Victoria.

Three of Yorta Yorta man Lee Joachim's children standing on the roots of an old river red gum in the Barmah Forest, Victoria, Yorta Yorta country.

Dead river red gums line a 'regulated' creek in outback NSW.

The Victoria River, now a 'channel' lined with dead river red gums.

The 'former' floodplain next to the Victoria 'channel'.

'Barmah Forest' by Lin Onus, 1994. © Lin Onus, licensed by VISCOPY, Australia, 2009.

Mary reveals how speaking for country is negotiated within intra-Indigenous politics. These responsibilities are known between traditional owner groups, who acknowledge and negotiate them with each other.

Importantly, these intergenerational responsibilities are references to a specific place: country. As Ben Smith has also described, these are knowledge systems embedded in country (Smith 2005, p. 8):

> … with innate ties to those persons who have a particular relationship with these areas, a relationship between people, land, 'law' and language … only those bodies of knowledge — and those people linked to them by innate 'spiritual' or substantial ties — are properly relevant to 'country' … it is only those persons who have such a substantive relationship with 'country', often based in having lived in a particular area, as well as being a member of a group whose forebears are from that area, who can 'talk for' the area in question.

In agreement with Steve Kinnane, Ben Smith has identified at least two criteria that must be met and publicly acknowledged for a person to be able to 'speak for country': inherited rights *and* a 'substantive relationship' with country. More than an inherited inter-generational responsibility, this authority is about deeply held relationships with country, expressed through religion, law, language and more. 'Speaking for country' involves listening to country, interpreting that knowledge, and communicating it to others. This authority is part of country.

When the MLDRIN delegates were working through their agreement with the environmental non-governmental organisations, the environmental advocates asked MLDRIN to provide them with a definition of 'speaking for country'. The MLDRIN delegates negotiated amongst themselves how to articulate this, and their explanation about their right to speak for country became part of the text of the agreement with the environmental groups (MLDRIN 2007b):

- The context for this right is the recognition that Traditional Owners have responsibilities towards their land and waters — the people, the land and water are one — and under traditional law plants, animals and places must be respected and cared for.
- Within Indigenous communities Traditional Owners hold the right to 'speak for country', a right that is inherited from ancestors and country, involving protocols and decision-making structures within clans/nations — which regulate who can speak and what they can speak about. In traditional law, Owners are associated with particular places, and through ancestors, an extended landscape. Knowledge and authority to speak for particular places or on particular issues is governed by protocols involving kinship, seniority, gender and so on.

- 'Speaking for country' within Indigenous communities is the Traditional Owners' birth right and responsibility to speak on the care of country and their inherent rights to land and water. It also refers to the unique right of Traditional Owners to speak for country in negotiation and consultation processes around land and water. This includes but is not limited to Native Title, Government and NGO policies and programs relevant to land and water, natural resource management, cultural heritage, sites management and protection including cultural sites and artefacts and sources of native food, State and National Park and Reserve management, water management including cultural flows, and engineering and other works that may impact on sites of significance.

In this explanation the MLDRIN delegates go from explaining what speaking for country means to making an argument for their involvement with the work of many other authorities. They relate their long-held authority to the modern work of natural resource management. While emphasising these relationships as distinctly Indigenous, their explanation is influenced by the language of natural resource management.

One of the main projects the MLDRIN delegates and the environmental non-government organisations they are involved with is the protection of the river red gums (see also NPA 2007). River red gums are still being logged on the river land immediate to the Murray because this land is managed as state forest by both the New South Wales and Victorian governments. MLDRIN and the environmental groups want to see these state forests become national parks, with joint management arrangements between the government and Indigenous people. To assist this, the National Parks Association of NSW challenged the legality of logging river red gums along the Murray, and a report by the Victorian Environment Assessment Council recommended that river red gum wetlands be protected (VEAC 2007). The Victorian Government has responded by creating a number of national parks along the river.

In negotiations about the river red gum campaign, MLDRIN executive officer Steven Ross has found himself constantly emphasising that the campaign is not just about trees. Steven has argued that the campaign also has to include water and the capacity building of the Indigenous nations (pers. comm. September 2007). Perhaps the isolation of the river red gums as a campaign issue by the environmental groups reflects the influence of modern thinking on the group, or, for people concerned about ecological relationships, it is a pragmatic decision about what makes an effective environmental campaign. The green advocates package their message to appeal to an audience used to representations of charismatic species unconnected to their ecologies; however, in doing so they appeal to ways of knowing the Murray that are not only green but also white. From the amodern perspective that Steven campaigns for, looking after the river red gums and the nation groups are inseparable. The capacity to speak for country is central to how best to care for country.

For the traditional owners, asserting their authority to speak for country is tricky because many other people have visions for what is best for country, including scientists and environmental groups. With their expertise, many scientists contend that they 'speak for' nature (Robin 2007, p. 9). Water scientist Peter Cullen celebrated the work of scientists, but also argued that they should not expect to have special standing to decide value questions for society, 'Their science needs to inform the debate, not replace the debate' (Cullen 2006). Likewise, thinking within one's discipline, or 'silo thinking', is critiqued in natural resource management literature as being too limited for the range of knowledge needed to address ecological devastation (see, for example, Bammer et al. 2005). It is my argument that all people, not just 'silo' thinkers, must first look to how our knowledge is constructed before we can make leaps to what the 'solutions' might be. Then we can reposition our knowledge as knowledge with authority amongst many other knowledges with authority.

Across Australia, traditional owners have used legal authorities to gain recognition for their authority to speak for country, as often cited in the *Mabo* case, but also notably articulated by the High Court in *Onus v Alcoa* (*Onus v Alcoa of Australia Ltd* [1981] 149 CLR 27). In this case the High Court compared traditional authority with an application made in an earlier case by an environmental group that had sought recognition for their authority to protect nature. *Onus v Alcoa* concerned a dispute over an aluminium smelter in south-west Victoria in Gunditjmara country; the dispute pivoting on the now familiar development versus culture binary. The Gunditjmara people argued that they had special interest in the care of this land and thus had special standing to oppose the smelter under heritage legislation. The smelter proponents argued that the Gunditjmara's special interest was insufficient as it was 'entirely emotional and intellectual', drawing a parallel to the reasoning of the High Court in dismissing an earlier case bought by environmental groups seeking special standing in order to protect part of Queensland (*Australian Conservation Foundation v The Commonwealth* (1980) 146 CLR 493). However, in *Onus v Alcoa* the High Court disagreed with the parallel drawn and determined instead that the Gunditjmara people were custodians of cultural sites in their country, according to their laws and customs. The justices recognised that the Gunditjmara claimants had sufficient interest to justify special standing in the court to speak for those sites.

Traditional owners hold and negotiate important relationships with country that are distinct from other groups in Australian society, distinct enough to be given special legal standing in *Onus v Alcoa*, and later recognised in *Mabo* as native title. Natural resource management focused on the best practice extensions of the latest western scientific knowledge fails to recognise the importance of these relationships. The nature-centred approach ensures that the dominant ideology of settler societies becomes the default standard for natural resource management, overriding Indigenous peoples' concerns about their responsibilities to country (Palmer 2006, p. 38). This is why the MLDRIN delegates value working with organisations such as CSIRO to influence their methodology, rather than having to negotiate that methodology after the project has started.

Debbie Rose theorises that damaged places are opportunities for dialogue with Indigenous people on ethical connections with country. Dialogue occurs where 'Entanglements give us grounds for action' (Rose 2004b, p. 22). To paraphrase her, we do not need to go out into the so-called wilderness to find common ground, we may do better to find it in scald areas, salinity, damaged sacred sites, and in parking lots, shopping malls and national parks (Rose 1999, p. 185). If we look to where the proliferation practices of the moderns have intensely upset their work of purification — where the binaries are so mixed up as to confound even the moderns' valiant efforts at separation — this unsettled ground opens up space for new knowledge to be heard.

Interestingly, the 2007 agreement between MLDRIN and the environmental non-government organisations contains two principles that emphasise the role of Indigenous science *and* western science in caring for country:

- that Indigenous science and Western science each have their own value and role in caring for country
- that knowledge and management work together — caring for country creates new knowledge and knowledge helps us better care for country

Here, both the traditional owners and the environmental groups explicitly acknowledge the transformative element of working together to produce 'new knowledge'. These principles move beyond debates about nature-centred approaches, but they were very hard to negotiate. According to Steven Ross, the environmental NGOs were concerned that calling traditional knowledge 'Indigenous science' would undermine the presumed universal authority of western science (pers. comm. October 2007). That the groups reached agreement reveals both the transformative effect of the dialogue and how strategic compromises are part of partnerships towards shared goals. Arguably, there are compromises for the delegates as well: labelling their holistic knowledge traditions as science, when the western scientific tradition has been dominated by separation thinking.

MLDRIN is a very good example of Indigenous people taking the initiative, working in a complex context where they both negotiate with governments and work to build the capacity of the Indigenous nations. They are not just thinking about river degradation and then articulating their concerns: they are doing things about it; they are responding with action. The mobilisation of MLDRIN by the traditional owners deliberately counters a key modern authority in natural resource management: the political neutrality assumed by scientists speaking for a constructed universal nature. The MLDRIN delegates are also seeking transformative relationships. The scale of work that they are engaged with requires dialogue that can engender transformations across all parties. In this translation and transformation context, the traditional owners have developed the terminology of 'cultural flows' to communicate their river restoration plans. Cultural flows might extend the principles of recognition, consent and continuity to other living beings in the Basin.

CHAPTER 6

'Deplete, destroy, depart?'

Don't touch the water.[1]

The expansion of modern water management in Australia was, and continues to be, an exercise by the moderns of their powerful knowledges. However, the power of modern knowledges has been their downfall. The moderns have been so successful at transforming ecological relationships that they have undone the life previously sustained by those relationships. In such a short time the life of the inland river country has been destroyed. In ruining this life support, the moderns negate their own dream. Without the river water, the Murray River System is not performing as the moderns had planned. Their power to intervene, direct, control and allocate nature is not producing the prosperity they expected. But what will their response be?

The traditional owners have responded with 'cultural flows', and a minimal definition of this new concept is that it is a way of returning water to the river country envisioned by the traditional owners. This is a conceptual tool to respond to loss, and to influence the responses of the moderns. Different representations of rivers and water carry different authorities; cultural flows is a representation that focuses on returning fresh water to country to revitalise and recharge webs of connectivity.

However, the influence of hyper-separated binaries profoundly confuses the message of cultural flows. To address this, the thinking that fosters hyper-separation must be addressed. All people, not just the traditional owners, need to engage with the power structures associated with the language of modern thought so as to engender more ethical relationships with the rivers. In particular, the moderns need to acknowledge the consequences of the assumptions they hold in modern water management. This has to happen first to create the space for dialogue with other knowledges.

---

1. Advice about the toxicity of Bottle Bend Lagoon near Mildura, given by a staff member of the Lower Murray Darling Catchment Management Association, to participants of CSIRO's 2nd National Indigenous Science and Research Roundtable, 6 November 2008.

We need to be open to an ecological dialogue to facilitate the flow of ideas and the creation of new knowledge for understanding our relationships with the rivers and our responses to river destruction — otherwise the moderns will continue to deplete, destroy and then depart elsewhere (a powerful phrase used by Grinde & Johansen 1995) to begin their destructive cycle again. Indeed, the federal government has dedicated $20 million to a task force assessing the Northern Territory's tropical rivers for large-scale agricultural expansion, partly in response to the devastation of agricultural land in the Murray–Darling Basin. Up north, the alarmed traditional owners are saying 'We don't want another Murray' (John Daly quoted in Hancock 2003). The task force has been created to avoid the mistakes of the Murray by investing in scientific expertise and a range of community consultation. However, it depends on what the 'mistakes' are perceived to be.

CULTURAL FLOWS AS A FORCE OF LIFE

The traditional owners speak about cultural flows as a way to return fresh water to the parched river ecologies. This water is described as rejuvenating life-sustaining connectivities. Yorta Yorta man Lee Joachim says that cultural flows will be different for each Indigenous nation and for individuals as well. Instead of just one understanding, the meaning of the cultural flow is particular to country and the people of that country. For Lee cultural flows are:

> ... a full flood that maintains all of the area, and is not just limited to a roadside or to a levy bank. It is actually flowing right throughout the system and ensuring that life is continuing within the system that it is supposed to nurture. [25 June 2004]

This flow would threaten homes, infrastructure, human lives, and livestock: ways of living designed around the containment of river water. But Lee has a counter-apocalyptic vision of the flood: the revitalisation of life. Lee positions the life brought by the river as more fundamental than all the other lives that are sustained in the river country, because Lee regards the river as central to ensuring the continuance of all that other life. Lee identifies the importance of what the river is doing. By attributing this agency to the river, Lee is arguing for the cessation of the 'management' of current natural resource management. Instead, cultural flows is a management that connects with and supports life and that sustains one's own life through connectivity. Indeed, Lee sees that the river is so resilient it can heal itself. The river can adapt to all the violence that has happened, and respond with new energy to continue to live and support life.

Lee has shown us how the language of connectivity is the language of cultural flows. All life is connected and other beings have agency. Humans are not dominating a separate and inferior nature. As Henry Atkinson has also said (Wagga Wagga, MLDRIN joint meeting with the CAC, 13 July 2005:

> Cultural flows are a natural flow which allows everything to grow. Cultural flows include your history and your culture.

Henry brings our attention to how history and culture are also rehydrated with cultural flows. In connectivity, country, life, culture and history are all intertwined. Again, Henry reveals to us how the arguments of the traditional owners pivot in connectivity. For Henry and Lee, their vision of cultural flows is about returning water in a variable way that supports the immense wetlands and river red gum forests of their Yorta Yorta heartland.

Out on the plains in western New South Wales, Mary Pappin spoke about her desire for the return of the floods that used to come and go:

> ... whether it just be the mouth of a little creek that floods up every now and then, traditionally that is where the resource would be growing after the flowing down into that area. [22 July 2004]

Mary knows that water is a variable presence. The plants that germinate after the flows come through are part of her inheritance, and she has intergenerational responsibilities to make sure these plants continue to live. Today the plants are absent, but their seeds may be surviving, waiting in the soil to germinate with the next flood waters. Mary's intimate knowledge of that presence, and her motivation to argue for those plants, is what she offers to cultural flows.

In central New South Wales, Wiradjuri Elder Tony Peachey grew up near the Macquarie River. Peachey says that the Macquarie used to have a good flow going at certain times of the year. However, since the dams, the weirs, the reservoirs, the pumps and the pipes, this flow has 'gradually dissipated':

> So unless you get that freshwater in to flush it out, the river itself just gets sick. That's as simple as it is; because the flows aren't there. [29 October 2004]

For Peachey, cultural flows are a flush of freshwater that enable the river to restore its own health, arresting the build up of weeds and the blue-green algae that multiply in still, nutrient-polluted water (Wagga Wagga, MLDRIN joint meeting with CAC, 13 July 2005). From his experience, Peachey knows that freshwater and river health go hand in hand.

Further south in the Coorong, the Ngarrindjeri people are distressed about salt levels in the water. With reduced freshwater flows from upstream, the shallow water in the Coorong has become saltier than seawater. This rupture of connectivity has diminished the life that previously thrived in the mixture of fresh and saltwater. Here, the Ngarrindjeri women can no longer find freshwater reeds and must head over to the Murray and the lakes next to the Murray to get the right reeds for basket weaving (Corowa 2006). Again, in this context, cultural flows are talked about as a return of the now-absent fresh water that sustained life. As Ngarrindjeri Elder Matt Rigney has said, 'Cultural flows — it's about the regeneration of life' (Wagga Wagga, MLDRIN joint meeting with CAC, 13 July 2005). Matt is interested in how cultural flows take care of 'our nurseries':

> What we talk about in terms of our nurseries are our swamps and wetlands. [Other] people see that as waste areas. Those waste areas are not viewed by those who are not in the know or don't want to be in the know, that they are the future ecosystems for our waterways.

The nurseries are where young lives are nurtured in the richly complex wetlands. Matt argues that the people who do not link freshwater ecologies with the future health of the rivers are either ignorant or are complicit in river destruction. The latter group diverts their attention away from the destruction that surrounds them in order to continue the activities that generate the destruction. Matt identifies the way in which many people regard wetlands as wastelands until they can be physically transformed into a 'useful' dam and paddock instead of a 'useless' messy swampy bog.

Matt has also talked about cultural flows in relation to taking care of the black swan. Matt is against current water management practices that send water down the river in the middle of summer when the black swans need that water to breed in the winter. This is a personal matter for him, as the black swan is one of his *ngatjis* — a Ngarrindjeri word for a kinship relationship with nature, known as a totem in other literature (see Rose 2003 et al., p. 10; Bell 1998, p. 199). Ngarrindjeri people inherit *ngatjis* from their parents, and they can be any of the local animals and insects. Matt has to care for the swan as a brother or sister. The swan is family. These are ethical relationships of interspecies kinship. Similar totemic systems across Australia, and around the world, provide Indigenous people with an intergenerational foundation for respectful and intimate relationships with nature.

Back upstream in the riverine plains, on the Edwards River anabranch of the Murray, Jeanette Crew talks about cultural flows as a caring agency. Jeanette said that when MLDRIN people speak about cultural flows they are:

> ... probably talking about cultural flows in terms of the availability of traditional foods and maybe looking after special sites, for instance, Bunyip holes, as you know are in the bottom of the river. If there is not enough water there for the Bunyip then perhaps it won't survive. [26 June 2004]

Jeanette links the physicality of the deep pools in the rivers to the survival of the Bunyip. The Bunyip is a mythical creature believed in by both Indigenous people and non-Indigenous people. Such creatures, and all the stories and sites of creative ancestral activity, are relationships that connect country with the Dreaming. In another example from South Australia where high cliffs line the Murray, Ngarrindjeri Elder Agnes Rigney told me how the 'monster type creature' the Mulyewongk lives in caves in the cliffs next to the river (see also Bell 1998, p. 38). The cave is near where the old Swan Reach mission used to be. When the old people went upstream fishing, they would always cross the river to the other side to avoid the cave. I asked Agnes if people still kept away from this spot:

> Oh yes ... That spot is still there on the river ... it is beautiful, it is white sand and the grass is just green. It is like a white beach, and the water then used to be clear, green, clear. You could see to the bottom of it back in them days, and it was a most beautiful spot and it was a forbidden one too. [21 July 2004]

The beauty and danger of these places is now covered by a slow flow of muddy regulated water that obliterates life. Ecological destruction destroys not just the plants and animals and their habitat, but the sentience of these places as well. Their ability to inspire fear and beauty is replaced with a bland utilitarian landscape, although the power of the Mulyewongk keeps on and the place is avoided by those in the know. Country continues. Agnes puts the past and the future together at the same time in the same place. She is using the language of continuity with country.

By turning the Murray River into a channel for consumptive water flows, the intricate intimacies of these places have been evened out, and the life of country has been diminished. This utilitarian role of the Murray is now failing the moderns. River destruction is the result of a model of water management based on water as a resource for exploitation; however, for the MLDRIN delegates cultural flows are not anti-development. Rather, in keeping with their holistic knowledge traditions, cultural flows include economic concerns. As Henry has said:

> When we are talking about a cultural flow, we mean the flow of water in the river system is completely natural, going with the season, and that cultural flow belongs to the Indigenous people, so they can get the benefit from the river system whether it is economic or to benefit the environment. In other words for

> the traditional owners of country to be able to use the water in any way they see fit. It is for looking after country the way the country should be looked after, in a natural way. [7 August 2004]

This is a different way of speaking about the river as an economic resource. A healthy river has economic and environmental benefits, but it is the natural river on which all this depends. There is no denied dependency here: the river country comes first. The rivers' seasons and cycles are to be respected, and in turn economic needs will be provided for as part of the reciprocal relationship.

In all these expressions of cultural flows the traditional owners do not just seek the return of water to country. At stake is their own agency as part of the cultural flow. This is part of their authority to speak and care for country. Thus, this cultural flow is also an expression of the ongoing responsibilities of traditional owners to their country; it is part of their active practice to care for country. Steven Ross has said that cultural flows will be the tangible realisation of the political activity of MLDRIN (pers. comm. 16 February 2006). When realised, they will be part of the political strength of the traditional owners, as the flows will recognise and recharge their identity and authority. A flow of water managed by government denies this meaning. As Wiradjuri Elder Ramsay Freeman asks rhetorically:

> What is the good of having Indigenous people in the country if someone else can come along and tell them what they should be doing in their own country? [27 June 2004]

In having 'their own' cultural flow, the traditional owners also wish to reduce the fraught and time-consuming exercise of translating what they want to do with the water and why they want to do it.

The traditional owners have developed the language of cultural flows to directly appeal to the importance of connections, countering knowledge that segregates and isolates. It is an evocative term, embedding culture and water together with movement through time and place. Importantly, the flow of water is uneven. Cultural flows recognise variability, and thus they leave and return. This is an ecological restoration approach that is not creating separated reserved lands or 'spaces for nature'. The territory of water is everywhere and everything. And within this territory the fresh water has its own agency and resilience in river restoration.

CULTURAL FLOWS AS A MODERN COMPROMISE

The traditional owners interpret cultural flows in the context of modern water management and water scarcity. Indeed, they would not need to make arguments about cultural flows if it was not for the excesses of modern water management. However, the modern water negotiation tables set a number of obstacles for the traditional owners in their translation task. When making arguments about cultural flows into this context, the MLDRIN delegates start speaking in reductionist terms, becoming trapped in modernity's language and also trapped by their own use of the term 'cultural'.

The traditional owners have coined the term 'cultural flows' to speak to policy-makers accustomed to the terminology of environmental flows. Ecologists have had some successes in arguing for environmental flows to restore river health. Initially environmental flows were envisaged as a one-off release of water from a dam. Today, environmental flows are described in policy as an ongoing responsibility for government to manage water as closely as possible to natural river hydrology (Thoms et al. 2004, p. 353; see also, MDBMC 2002, p. 23). Hydrologists, geomorphologists and ecologists now research how these planned water releases relate to river hydraulics, the speed of the river flow and habitat (Dyer et al. 2003, p. 7; Thoms et al. 2004, p. 365).

In the rhetoric of modern water management, the government draws on the expertise of scientists to determine the best-practice management of the rivers. The traditional owners insert themselves into this space by using the word 'culture', and thus make a point about having different concerns that are not encompassed by 'universal' science. They assert their distinct Indigenous identity and political status with culture. Further, the delegates direct attention to the influence of culture in their knowledge frameworks; in contrast, this is something that is absent from policy literature on water and natural resource management. However, by lobbying for culture the traditional owners risk buying into the modern baggage that comes with this term. Indigenous people are stereotyped by modern knowledge as cultural beings: they are not rational thinkers, they are not capable of objectivity, they do not 'use' ecological resources, and they do not have economic agency. As the aluminium smelter proponents in *Onus v Alcoa* argued, their interests in country are 'entirely intellectual or emotional'. But all people have culture, westerners included. Indeed, the traditional owners have to emphasise cultural diversity because their experience of the modern universals is that they are not actually universal at all. Furthermore, that they are partial, fragmented and driven by linear trajectories of progress.

Because of the problems with the word 'culture', the MLDRIN delegates sometimes substitute the phrase 'Indigenous water allocations' for cultural flows. This language does not come with the trappings of primitivism, but it is also less evocative of the rivers. Indeed, it evokes the storage and control of water, and thus offers implicit support for the conceptual foundations of modern water management.

On the MLDRIN website the following information about 'Indigenous water allocations' is posted (MLDRIN 2007a; see also Ross 2006–07):

> MLDRIN has as its core objective in the coming years the establishment and implementation of Indigenous Water Allocations. This water would be used for a cultural purpose; this means a water allocation would be used for whatever purpose the recipient Indigenous Nation deems culturally appropriate.
>
> Through a negotiation process with each other and with the relevant jurisdiction, the traditional owners would decide where and when water would be released. The water could be used for cultural economic purposes such as to water a native food source or medicinal plant source, enable breeding of native animals through appropriate flooding of wetlands or other floodplain ecological system or send water to an important spiritual or cultural site.
>
> An Indigenous Water Allocation will support the continuation of our cultural practices and could have significant environmental outcomes.

In this text, it is argued that an Indigenous water allocation *is* cultural, and culture is defined as whatever the traditional owners think is culturally appropriate. This is a tricky negotiation: appealing to the 'cultural box' within natural resource management, and then asserting that this box is unlimited. The formal language that identifies and translates culture reduces the arguments made earlier about the regeneration of life in the river country and weakens the vivid passion of the delegates when they speak more freely about cultural flows.

The language switches the MLDRIN delegates make between 'Indigenous water allocations' and 'cultural flows' is evidence of how influential modern water management thinking filters into the delegates' own thinking in their engagements with governments. But these language switches are also political. Politics is an inextricable component of the negotiations, and thus different language is used. This is illustrated by MLDRIN's strategic plan. The delegates developed two versions of their strategic plan: one for themselves and one that is publicly available. Rhetoric is separate to practice. The delegates use the language of 'allocations', of water as a mute resource, in order to create more space for their own water 'management'. They use the authority of modern language to extend their authority: if humans are directing the allocation of water they can also participate in this activity. But they end up speaking a language that fits more easily within the knowledge framework of modern water management.

The complexity of this translation context, and the difficult choices that the traditional owners are presented with, can be illustrated further by comparing how the traditional owners relate cultural flows to environmental flows. Ancestral beings and cultural living are not part of the matrix examined by the hydrologists, ecologists and geomorphologists when deciding which wetlands to sustain with an environmental flow. Nor do the managers of environmental flows consider the ramifications of whose traditional country will miss out when they decide to water a wetland. Thus the traditional owners cannot rely on environmental flows to be responsive to their relationships with country. Indeed, they are keen to emphasise a distinction between cultural flows and environmental flows to policy-makers. However, articulating this distinction at the water management table is limited when meaning is confounded by modern thinking and political choices to use and subvert that modern thinking.

In July 2005 the MLDRIN delegates met with the CAC in Wagga Wagga, New South Wales, to discuss the meaning of cultural flows. At this meeting Steven Ross emphasised a synergy existing between cultural and environmental flows (Wagga Wagga, MLDRIN joint meeting with CAC, 13 July 2005). Matt Rigney agreed and said that the traditional owners needed environmental flows to take care of the birds, animals and fish. The development of environmental flows in government policy marks an important step towards the amodern arguments made by the traditional owners, and the delegates acknowledge this. If the river were flowing well, as it used to, naturally, then the traditional owners would not need to make arguments about cultural flows. But in the context of water scarcity and the dominance of consumptive uses of water in water management, the traditional owners continue to argue for cultural flows. As we have seen with The Living Murray, at this stage environmental flows only have a very limited size and territory. Indeed, as at September 2007 less than 1 per cent of total water currently available in the large storages has been allocated by governments for 'environmental use' (MDBC 2007f, p. 4). Mary Pappin has spoken about how she appreciates environmental flows, but remains concerned that these flows are small and selective.

While environmental flows are a recognition by the policy-makers that the river has its own water needs, environmental flows are still a 'human-made' allocation of 'environmental water'. Environmental flows are constrained by the natural resource management framework that prioritises economic values, perceives ecology and economy as oppositional goals, and presumes that humans are the only ones who have agency to intervene and direct river flows. The ongoing mentality that the river is a competitor for water is evident when environmental flows are illegally diverted by landholders. Dramatic jumps in water prices meant that it became cheaper for irrigators to steal water and pay the fine than to pay for water up front (Jenkin 2007a, p. 3). This is also evident in the prioritisation of water allocations during times of water scarcity, when environmental flows are placed second behind consumptive water uses (Grafton & O'Connell 2008, pp. 74–5).

Significantly, another reason why traditional owners are keen to establish a distinction between environmental and cultural flows is the negative effects of environmental flows. Environmental flows are complex undertakings, have had a mixed history of implementation, and are only recently being researched for their effectiveness (Smith 2001, p. 296; Blanch 1999; Harman & Stewardson 2005, p. 113). Environmental flows might 'top-up' high flows from unregulated tributaries, to create a larger flooding event (Harman and Stewardson 2005, p. 113). Environmental flows can also be managed to drown out structures such as weirs so that migrating fish can pass (Thoms et al. 2004, p. 351). For the Darling River, the concept of environmental flows has to include the importance of periods of no flow, when the river runs dry (ibid., p. 365). If environmental flows are to mimic variability, then it is not a matter of simply releasing water in a steady flow, but storing water for timed strategic releases or pulses. However, a cold environmental flow released from the bottom of a dam would be better described as thermal pollution. In addition to practical complexities in the implementation of environmental flows, the (mis)labelling of water for other purposes as 'environmental' contributes to their bad press. The unseasonal movement of water along ephemeral river beds to downstream irrigators, including the flooding of the Barmah–Millewa forest, has been called an environmental flow. Crisis measures to flush water down a river in response to an outbreak of blue-green algae have also been defined as environmental flows (Smith 2001, p. 296).

Both Lee and Henry have had bad experiences with environmental flows in the heartland of their country, the Barmah–Millewa wetlands. 'Rain rejection flows' flood this wetland with too many small floods at the wrong time of year, while the higher country is left dry. Henry spoke at the meeting with the CAC about how environmental flows had killed life in his country:

> The environmental flow denies the Indigenous people access to hunt and gather their natural foods, plants and medicines. [7 August 2004]

For Henry, environmental flows are a continuation of the destructive mismanagement of his country. Lee pessimistically regards an environmental flow as a flow that only farmers and irrigators gain from, as part of a knowledge framework within which the river is regarded as a resource for human consumption. Indeed, the Barmah–Millewa wetlands are described by irrigators and river operators as a 'choke'. Here, the tight twists of the Murray's path slow down the river's flow, or, as modern water managers have constructed it, 'choke' the delivery of water to downstream irrigators. However, since these interviews with Henry and Lee, a bigger environmental flow was released under The Living Murray to flood the Barmah–Millewa forest. I saw Lee at this time, in Echuca in October 2005. He was happy that a larger flood of water was spreading through the forest, but was still concerned that the flood did not go further into the bush. Lee remains worried about the way the monitoring of environmental flows is focused on fish and/or bird breeding, and not on encompassing all the life of country.

Part of the MLDRIN delegates' argument for cultural flows relates to their political relationship and responsibilities with country. Mary Pappin made the distinction that environmental flows are controlled by the 'white man', and do not take care of country as it should be 'culturally'. When the traditional owners speak about caring for country, they are speaking about a reciprocal relationship whereby country is also taking care of them. This is an authority they carry to their negotiations over water; however, this communication can become blighted when translated into natural resource management terms. This is evident in what Steven Ross has said about cultural flows (Corowa 2006):

> The outcome might be similar to an environmental flow, and it could be used to supplement an environmental flow, but it's who has control of it, and who has control of the mechanism, and the timing, and where it goes.

For people familiar with connectivity thinking, this comment can be understood in context of responsibilities to country, a management that is not centred on human dominance of an inferior nature. However, Steven's comment could be easily taken to describe just that, as it would be in a natural resource management context. As MLDRIN executive officer, Steven finds his role as an intermediary involves codifying his language for both sides, the traditional owners and government (pers. comm. October 2007). It is important to remember the highly charged political context of this dialogue. Ramsay Freeman has vividly summed up the difficult situation that the traditional owners face in their strategic engagements with government: 'You can't walk along a barbed wire fence with one foot in each paddock' (pers. comm. October 2007).

Cultural flows are not a cultural copy of environmental flows, one taking care of cultural values, and the other taking care of environmental values. Cultural flows are part of the MLDRIN delegates' critique of the dominant paradigms in water management today. This is a bigger vision of the cultural flow, which demands a philosophical shift in natural resource management. This bigger vision is perhaps not being communicated by Indigenous water allocations.

In practice, the realisation of cultural flows as a return to how the rivers were is an extremely challenging ambition of the traditional owners. The government has not accepted its own commissioned scientific research that recommends the return of 4000 gigalitres for a chance at restoring river health. The political consequences of reducing water allocations are too much of a challenge for governments. More significantly, with ongoing drought the government is not in a position to act on those recommendations. In September 2006, water storages in the Murray River System totalled 3350 gigalitres, one year later they total 2130 gigalitres (MDBC 2007f, p. 3). Parched country meant that there was very little run-off into the storages from the 2007 autumn rains.

The realisation of any sort of cultural flow is now much more dependent on future rainfall than the negotiation skills of the MLDRIN delegates and the capacity of policy-makers to hear these arguments. If there is rain, and the storage levels rise, then the delegates may get their allocation of water for cultural flows. However, even though the traditional owners wish to recharge connectivity, it is likely that in the short term cultural flows will be realised from within the Murray River System. The cultural flows will have to be planned with respect to size and timing, moved through canals and channels, as regulated by weirs and locks. Pumps and pipes will be needed to breach the ruptured connection between the former wetlands now isolated from the rivers. And the hydrologists, ecologists and geomorphologists will provide valuable technical advice, gleaned from their experience with environmental flows. Hopefully for the traditional owners, and for all of us, even this cultural flow will be part of the many other catalysts that are already contributing to demands for the transformation of water management.

ECOLOGY/ECONOMY AND CULTURAL ECONOMY

Identifying how to support economic outcomes without also harming the rivers is key to the future of the river country. The traditional owners work to counter the hyper-separated ecology/economy binary with their ideas about 'cultural economy'. This cultural economy is intended as a translation tool to counter modern thinking, including the commodification of nature as natural resource, to inform governments' responses to river degradation, and as their own vehicle to sustain their relationships with country.

Modern knowledge situates economy and ecology in oppositional relationships, presenting us with a false choice: one must be sacrificed for the other. This construction informs understandings of how ecological devastation happened: river health was sacrificed for the agricultural industry. This construction also informs responses to ecological destruction: environmental flows will be at the cost of the livelihoods of water entitlement holders. The notion that ecology is a resource that must be sacrificed for human use is central to The Living Murray's perception of the Murray. In this policy, the Murray is to be managed to ensure a 'healthy working river', and, as cited earlier, this is defined as (MDBMC 2002, p. 47):

> ... one that is managed to provide a compromise, agreed to by the community, between the condition of the river and the level of human use.

It is a definition based on compromise, pivoting on the perceived oppositional relationship between ecology and economy, and it continues the imagery of human relationships with nature as one of 'use'. In public debates about water in Australia, the ongoing drought is leading to a shift away from hyper-separated ecology/economy binaries. The theorising and practice around ecologically sustainable development is also influencing this shift.

However, modern assumptions about nature as natural resources for human use continue to underlie government responses to ecological devastation, including policies of ecologically sustainable development. For example, governments emphasise water trading as a useful response to river degradation as they believe that the market will take care of better water use in the Murray–Darling Basin (Federal Liberal Party Senator Bill Heffernan, speaking in McMullen 2007). In this market-driven economy the dramatic increases in the price of water would reduce inefficient water practices and ensure that the use of water is maximised for economic productivity. Thus we can leave it up to the market to ensure that furrow cotton and paddy rice will not be part of the future of the Basin (McMullen 2007). However, the market may also decide that expensive wines are the most efficient economic product, in which case political arguments about the Murray–Darling Basin as Australia's breadbasket would no longer stand valid. Arguably, capitalist commodification is so influential in terms of its capacity to incorporate nature that economists and politicians need to recognise market failure and address it with market or legislative tools. However, this thinking stays within the tools of capitalist commodification and does not address the harmful ecology/economy hyper-separation. It is not the economy or economic motivations that have so profoundly transformed nature, but western rationalist economics that has lead to the commodification of water as a resource for exploitation (see, for example, Main

2005). Indeed, the market will not do anything to address the over-allocation of water; it will just ensure a more efficient consumptive use of water.

The arguments made by the traditional owners about cultural economy seek to counter harmful modern binaries by bringing connectivity into the foreground. The term cultural economy (or customary economy) has been used elsewhere in Australia to describe the subsistence economy of the traditional owners (Povenelli 1993, p. 5; Altman 2003a, p. 2). Here, the MLDRIN delegates use cultural economy to express themes of ecological restoration and repair, using the logic of holism to connect ecology, culture and economy (see, for example, Ross 2006–07, p. 18). Mutti Mutti Elder Jeanette Crew talked to me about how Wamba Wamba women, her close relatives, want to revive the art of making woven grass baskets and trade them as part of their cultural economy. Jeanette said that the way the water is managed today is 'interfering with our cultural economy'. For the grasses to grow the seasonal flood waters need to return to the swamps in the Werai forest near Deniliquin. In this industry, Jeanette has identified how traditional owners can care for country, meet responsibilities to their ancestral creators and future generations *and* earn money.

Wiradjuri Elder Tony Peachey makes me think of cultural economy when he speaks about the possibility of tours by traditional owners in the Macquarie Marshes Nature Reserve. The NSW National Parks and Wildlife Service staff negotiate with Peachey and other Wiradjuri people to be shown more cultural heritage sites in the Macquarie Marshes as part of the Service's management of the reserve. Peachey is cautious about sharing his knowledge. He worries that the parks people will want to include these sites in their own tours, and this increased visitation will be out of the traditional owners' care. Peachey proposes instead that the tours be run by the local traditional owners so that they can directly manage their responsibilities to the sites. This role also recognises their authority to interpret and speak for country, and would provide them with regular opportunities to be out at the marshes. Peachey told me about the lads from the nearby towns of Warren and Quambone who go to the marshes pig hunting and fishing. Peachey wants to encourage this relationship held between young people and country by creating job opportunities for them at the marshes.

Economist Jon Altman has extensively researched and lobbied for ways in which government policy can facilitate positive outcomes for culture, economy and ecology. Jon Altman has argued that there are clear policy synergies that governments can achieve by supporting Indigenous people in environmental jobs on country. The governments have an opportunity to match the traditional owners' aspirations to care for country with public policy concerns about Indigenous wellbeing and healthy landscapes (Altman 2003b). Indeed, in the 2007/08 budget the Federal Government boosted the funding of the Indigenous Protected Areas program, a program that provides money for environmental jobs on Indigenous owned land (Turnbull 2007). Jon Altman's argument relates to the large land holdings of Indigenous people in northern Australia; however, his policy arguments can be applied more broadly. These are similar to the arguments the MLDRIN delegates keep making when they point out how multiple positive beneficial outcomes can be achieved by thinking about connections rather than separations.

The Indigenous knowledge traditions behind cultural economy provide a different methodology for how to earn a livelihood from country. However, by using the term 'cultural' in front of 'economy' the traditional owners are again appealing to government notions of Indigenous people as 'cultural', and thus run the same series of miscommunication risks as with cultural flows. The phrase 'cultural economy' suggests economic activities that reinforce Indigenous cultural identities, such as 'eco-cultural services'. This can be compared with, for example, water trading on the free market. Indigenous holistic knowledge keeps bumping into these odd and contradictory binds: that what people want for the good of the whole system (including themselves) gets classed as culture because that is the only conceptual space available to them. By speaking to modern frameworks, the connections the MLDRIN delegates make between economy, ecology and culture are not possible. Indeed, from the perspective of ecology/economy as a hyper-separated binary, when the traditional owners argue that cultural flows include economic concerns, this argument is cast as antithetical to their distress about river health. In modern thinking Indigenous people are either premoderns or primitives and do not have economic agency or commercial concerns. Or, if they do pursue economic goals, then they are no longer authentically Indigenous.

Indigenous lawyer Jason Behrendt and non-Indigenous lawyer Peter Thompson have argued that culture and economy be addressed by separate policies. They recommend that cultural flows have to be distanced from the commodification of water as a tradeable right, and be legislated as a non-tradeable interest (Behrendt & Thompson 2003, p. 59). They suggest separate policy tools for Indigenous people to have the opportunity to use water as an economic asset (in particular, the NSW Aboriginal Water Trust; see pp. 71–2). Jason Behrendt and Peter Thompson draw on the modern preference for separation and protection in these arguments. Indeed, non-tradeable cultural flows could ensure greater protection of cultural flows for future generations. Yet trading cultural flows on the market is not subject to the same concerns about intergenerational equity as selling land. Water entitlements are annually renewable, unlike the sale of land. However, a scenario where water is sold *every* year would raise these concerns.

If the government could see past modern restrictions about what is and is not culture, and allocate MLDRIN the opportunity to manage water that they were able to trade, it would provide the traditional owners greater opportunity to take care of country and build their capacity. This would allow Indigenous identity and economic development to be seen as complementary and interrelated, not contradictory and oppositional (Ridgeway 2005).

The importance of economic matters for the traditional owners has grown rapidly in response to the dominance of capitalist values and systems since colonialism. Indeed, Jon Altman makes the link between Indigenous peoples' historical alienation from their economic assets to their contemporary economic marginalisation (Altman 2005, p. 44). He further argues that the recognition of emerging property rights for Indigenous people — such as that over water — must break with past practice and include their commercial potential so that Indigenous communities can establish themselves as viable economic entities (Altman 2004, p. 29). The historical irony of debates about whether Indigenous people have economic rights or not is pointed out by Barkandji Elder Junette Mitchell when she speaks about how quickly circumstances have changed for Indigenous people. Not long ago the Indigenous people showed the explorers and settlers where the water was, which led to the progressive dispossession of Indigenous peoples' land and economies. Today, Indigenous people cannot use water directly from the river, and instead pay water rates to access town water. The traditional owners' experience with colonisation makes engagements with the market economy not just a matter of choice, but a matter of necessity. Mutti Mutti Elder Mary Pappin brings the agency of carp into this debate as an active agent of dispossession. She looks about her country and sees how the 'white man' took all the good bits of land and the carp took all the good bits of the river, 'Especially our favourite fishing spots!'

While it may be easier or more desirable for settler Australians to identify the prejudicial violence of modern knowledge against Aboriginal people as historical, and participate in an 'ask forgiveness and then forget' reconciliation, abuses to Indigenous peoples' identities continue today (for example, see Grinde & Johansen 1995, p.18, on reconciliation with North American First Nation peoples). As cultural theorist Stephen Muecke describes (Muecke 2004, p. 6):

> We are careful now, in our relativistic way, not to designate the whole of European civilization as superior to the one it would largely displace, but we are happy to allow one of its minor but key concepts — the modern — to continue on a search-and-destroy mission.

Natural resource management and water management are not excluded from this 'search-and-destroy' work. I have shown this by examining the complex translations and strategic political choices Indigenous people make when negotiating their rights and interests with governments. An examination of the assumptions behind modern representations such as 'nature', 'culture', 'economy' and 'nation' continue to be important work relevant for settler Australians in decolonising relationships with Indigenous people today.

ACKNOWLEDGING ECOCIDE

There is something we must do before we can begin to do anything about river destruction. We have to look directly at 'ecocide' and acknowledge that it has happened. Without this knowledge we cannot make the decisions about how to respond.

Around the world human-made environmental destruction, or 'ecocide', has destroyed life-sustaining wetlands and freshwater ecologies. Indigenous people across the globe have had their way of life threatened by this destruction of their lands and waters. For example, in 1986 in eastern Canada, pregnant women — including women from the Mohawk Nation — were told by their state government not to eat the fish from the Saint Lawrence River as the fish were contaminated by carcinogenic polychlorinated biphenyls that had been dumped into the rivers by decision-makers in heavy industry. In 1990, this warning was extended to all people (Grinde & Johansen 1995, p. 189). Pronouncements by governments to their citizens to stop drinking the water and eating the fish are profound. They require that the people find different water and different fish elsewhere. The pronouncements formally identify that country cannot support human life. This is not a sustainable practice.

It is only possible for the profundity of such pronouncements to be invisible if ecocide goes unacknowledged. Indeed, ignoring ecocide is an important part of the power of the moderns. As Bruno Latour has argued, the moderns can only exercise power by ignoring the proliferation of what results from their work of purification. In the water administration area known as the Murray–Darling Basin, this means that the modern water managers are only able to keep working with water as a resource separate from land and life if they can ignore the consequences. They must ignore the desiccation of river life, the loss of river variability, the rising watertable, and so on. These changes are a big message about river health that the moderns fail to hear. To hear this message is to unsettle the foundations of their management approach. In contrast, the traditional owners do not face this challenge. They are able to acknowledge the loss of river life without unsettling the foundations of their knowledge. Indeed, it is connectivity that makes them alert to what is going on, and their belief in the resilience of the river country that urges them into action to address this loss.

Yorta Yorta woman Monica Morgan expresses a passion for acknowledging loss as the first step in restoring the life of country. Over her lifetime Monica has witnessed the loss of swans, pelicans, leeches and duckweed in her river country, and the loss of places where the water was so clear you could see vegetation growing on the bottom of the river. She speaks about the loss of the life of country with the losses for the people of country:

> If we're seeing this loss, we need to acknowledge the loss, see what the loss is, really engage with Indigenous peoples. Be able to relate history, relate connection to country, do that whole mapping of what the knowledge was, and what is still there, and what's the loss. [1 July 2004]

Monica illustrates loss in Yorta Yorta country with reference to the painting of the Barmah Forest by Yorta Yorta man the late Lin Onus (see following image). This is a beautiful landscape picture of tall river red gums and their reflection in the lake. Four sections of the picture are dislodged, cut out in the shape of jigsaw pieces. Monica interprets the missing jigsaw pieces as the magpie geese, the native trout, the brolgas and others that are now absent from the Barmah Forest:

> We need to see what would make that piece fit back in again. There are a lot of things that need to be rectified, and that's what needs to be done ... If we don't start returning the river and the proper watering regimes ... we won't see those parts of that jigsaw. It will actually become more of a disintegrated picture. [1 July 2004]

Monica continues with what the response to loss might be: cleaning up the rivers and improving the water quality. Monica moves from acknowledging loss to developing a plan of action, relying on the resilience of the river and other agencies. She describes how ecological devastation is inseparable from what remains and that a productive response to this loss can move through time to heal the past and build a future. Cultural theorists David Eng and David Kazanjian also argue that loss is not only a negative experience, but has creative potential that works on the tension between past and future in interesting ways. They write that 'the past remains steadfastly alive for the political work of the present' (Eng & Kazanjian 2003, pp. 1–2, 5). The loss becomes a resource for the present, and the future is to recover or redeem what has been lost (Butler 2003, p. 467). Indeed, continuity — which can be thought of

'Barmah Forest' by Lin Onus, 1994. © Lin Onus, licensed by VISCO-PY, Australia, 2009. (See also colour image, between pp. 112–13.)

as connectivity across time — will always accommodate both loss and creativity. That is the meaning of flux and flow, of how change, including death, is always part of living. However, the losses that are occurring are on such a catastrophic scale that they are signalling the death of connectivity with profound consequences for the future. These extended ramifications are what Monica identifies when she worries that the wetlands will become more of a disintegrated picture.

The traditional owners look at the fish on the end of their fishing line. They cannot always eat what is there. Junette talked about how she used to walk in the clear water looking for shrimps and yabbies as a kid, before the weir was put in at Menindee that transformed ephemeral lakes into permanent water storages. Now, she says:

> It is too dirty. And all that blue-green algae on top of the water. The fish are sort of green and slimy looking. They have sores on them. It puts me off eating the fish. It is not the same as when I was a kid … Since they put the weirs in. That is what kills everything. [20 July 2004]

6. 'DEPLETE, DESTROY, DEPART?'

The diminishment of the life of country is not just affecting the traditional owners; this life is something all Australians share. Historian Paul Sinclair argues that acknowledging loss is important work for all people living in Murray River country. Settlers need to recognise what is being denied as the next necessary step in river management (Sinclair 2001, p. 234):

> Settler Australians need to understand and mourn the immense losses they have inflicted on the river. They need to recognise the stories within their culture that can help them imagine a new future for the Murray.

Sinclair analyses tourism brochures that popularise the Murray as an escape into beautiful wildlife, rather than admitting the river's flow is highly regulated (ibid., pp. 233–4). This advertising continues today despite the ever-worsening drought. Newspaper articles promote the beauty of the 'Mighty Murray's majesty' to encourage the tourism industry (Anon 2007b, p. 36). Tourism is one of many economies that is declining because of river degradation. Houseboat operators report that their bookings are down 35 per cent because of the perception of a dying Murray now dominant in the national media. The operators insist that the water is still there — the regulated flow means that there is always some water that the boats can float on (Jenkin 2007b, p. 56). But will the tourists be fooled that this flow is the majestic mighty Murray?

In the media the damage to the Murray and Darling rivers is symbolised by the death of the iconic red river gum forests. Aerial shots of country show the extent of the devastation: the tall trees are dead but still standing on the parched earth. The trees are adapted to the variable regime of floods and droughts, and characteristically line the banks of the ephemeral inland rivers and creeks. But the 'regulated' drought combined with the climatic drought has pushed these trees beyond their survival capacity (see photograph A, p. 138). These scenes are repeated wherever this modern water management has been applied. The riverbank of the Victoria River is now a gravel divider, channelling the water to consumptive users downstream and causing desertification of the country this water used to flood. The regular flow of water has killed the red gums along its banks (see photograph B, p. 139). Adjacent to the channel the trees on the former floodplains that once moved in their dance of water and light are now caught in a freeze frame (see photograph C, p. 139). A dead tree provides habitat for animals, and in Wamba Wamba country the hollows are good for the *murrupmaginnie* and the *nyutha*, the creatures Jeanette Crew talked about in Chapter 4. But a desiccated floodplain of dead trees is not feeding energy to other life in connectivity.

Dead river red gums line a 'regulated' creek in outback New South Wales. (See also colour plate, between pp. 112–13.)

Aboriginal people believe they have always been a part of country. Country and people of country have been co-present since their creation in the Dreaming. Knowledge of co-dependent life is what the traditional owners want to share with all people, as we all share the same earth. The irony is that modern knowledge frameworks confound expressions of loss by Indigenous people. The positioning of loss and culture in the tradition/change dualism creates a conundrum for the MLDRIN delegates. It is because of the ecological loss, and its related impacts on their culture, that they are speaking out about river degradation. However, to gain a place at the water negotiation table as traditional owners, they are encouraged to make assertions against cultural loss in order to prove their 'authenticity' as Indigenous people. But the models of hierarchical civilisation and progressive reason are not the only way to interpret loss. Where the traditional owners identify the loss of cultural living, this is not an inevitable loss of Indigenous identity in exchange for a superior modern identity but the diminishment of all life, their life included.

6. 'DEPLETE, DESTROY, DEPART?'

The Victoria River, now a 'channel' lined with dead river red gums. (See also colour plate, between pp. 112–13.)

The 'former' floodplain next to the Victoria 'channel'. (See also colour plate, between pp. 112–13.)

Whilst I have focused on the river, similar devastating processes are occurring across a wide range of species and habitats. As I write it is springtime and the annual migration of the bogong moths is under way. They have descended upon Canberra in their millions. At this time of year the moths fly to the Snowy Mountains to aestivate, which is the summer version of hibernating. Famously, Aboriginal people used to walk long distances to the mountains to feast on the bogong moths massed in the crevices of caves (Flood 1976, p. 43). Bogong moth recipes now appear in the newspapers at this time of year as bush tucker, but we are warned not to eat too many. There is no possibility for feasting. The bogongs carry concentrations of arsenic, which is present in the soil of their larval pasturelands, the agricultural fields where arsenic has been sprayed (Australian Museum 2007). This arsenic is more than a threat to cultural living; it is another example of the diminishment of life.

The loss of variable freshwater flows in the inland river country is not just the loss of one link in connectivity, but the loss of a key connecting life force. Fresh water rejuvenates life. When the rain comes it soaks into the dry earth and when that earth is soaked the water runs off into creeks, streams and rivers. The flooding rivers spread fertile nutrients over the land. Kangaroos, emus, dingoes, birds and bats move large distances to avoid the floodwaters, while skinks, geckoes and koalas seek local refuge in trees and on higher ground (Young et al. 2001a, p. 85). Snakes take advantage of the mice and lizards marooned by floodwaters and are able to raise extra young at this time. Arid frog species rapidly develop to take advantage of the temporary waterholes in puddles after heavy rain. With very heavy rainfall and prolonged floodwaters the burrowing giant banjo frog emerges from underneath the hardened clay (Scott 2001, pp. 283, 280). After the floods, small aquatic animals flourish, and breeding waterbirds feed on the fresh food (Young et al. 2001a, p. 86). With receding floodwaters, plants send up shoots, set fruit and seed, and all mammals, humans included, are attracted to the feed. Floodwaters celebrate and sustain life. This is what we are losing when we transform complex freshwater ecologies into channels that deliver water for consumptive use.

I wonder at what point do those of us heavily influenced by modernity stop and think about what we are losing here. We have already lost 90 per cent of the wetlands in the Murray–Darling Basin. Monica Morgan expressed her frustration about the delay in action. The white fella science has identified the damage, but in terms of reducing the consumptive uses of water, not much has happened. If ecocide remains unacknowledged, then the easiest path for institutions such as governments is to continue the *status quo* of deplete, destroy, depart. As Daniel Connell said about the meagre environmental flows targeted in The Living Murray, this is 'a philosophy of despair' (Connell 2005b, p. 285). We need to see the destruction that has occurred, to come to terms with what has happened before connectivity is destroyed. It is not the carp, the weirs, the salt, or the pesticides that we should be worrying about. It is the violence of modern thought and action that is the fundamental threat to the river country and to all the people of that country.

The Indigenous people of Japan, the Ainu, have written a book: *Our land was a forest: An Ainu memoir* (Kayano 1994). Who amongst us wants to read *Our country was a wetland: A Yorta Yorta memoir*?

REPRESENTATIONS AND AUTHORITY

The world of representations created by modern knowledge carries so much authority that the representations have replaced the real world. This is explicitly illustrated by the Running the River arrangement constructed up by Murray River Water, as introduced in Chapter 1. In Running the River the distinction between the real and the represented is offered as unproblematic. The system has become the river for the river operators whose work does not involve getting their hands dirty with muddy salty water. Scientific data is collected automatically by machines sited along the river and sent daily to Canberra to ensure best practice.

At the start of this book I identified how the moderns emphasise two narratives of the Murray River, the first describes the rivers as a spiritual homeland, and the second an economic resource. This modern view of country is illustrated by putting two perceptions of Pink Lake into an oppositional relationship (see photographs p.142).

Matt Rigney explained to me that a battle was held on this site between the Ngarrindjeri people and the Ngarkat people who lived to the west. The pink is the blood of the warriors who were killed in the battle. This story of an ancestral homeland is easily placed in an oppositional relationship to the interpretive sign at Pink Lake that espouses a reductionist and utilitarian perspective This sign reads:

PINK LAKE
Pink colouring caused by
a combination of *Dunalialla
Salina* and *Halo Bacteria
Dunaliallla Salina* is
processed and used as
a natural food colour
and as an anti-oxidant.

Above: The Pink Lake, South Australia. (See also colour plate, between pp. 112–13.) Right: Signage at Pink Lake, South Australia.

Instead of developing and entrenching these two perceptions as oppositional irreconcilable narratives, what I have done in this book is show that their separation is an artificial modern construction, unhelpful in fostering dialogue between Indigenous and non-Indigenous knowledge and a contributor to contemporary ecocide. Binarised debates about Indigenous and western knowledge, or local and universal knowledge, only serves to reinforce those categories. The real issue is to engage with the dynamism of diverse knowledge (Sillitoe 2007, p. 9). By having multiple perspectives to draw on, we all become more resilient and able to change and address new circumstances (Grinde & Johansen 1995, p. 273).

Luckily, we are not actually living in modernity. Modernity is a state of mind, not a stage in history. Our contemporary world is a lot more diverse, and modern knowledge is just one of many ways of interpreting the world. Shifts in global climate, the destruction of large rivers, and other environmental catastrophes have shaken modern knowledge. But this knowledge does not have to be thrown out in a postmodern apocalypse. We need to acknowledge emotion, religion and intuition as inherent to science, reason and rationality. Anthropologist Diane Bell learnt about the importance of interpreting signs and feelings from her work with Ngarrindjeri people. She wrote about the dilemma of this knowledge for westerners (Bell 1998, pp. 224–5):

> The notion that one could know by feeling goes against the rationalist grain. If it can't be proved or disproved, can it [be] considered knowledge? As heirs of the scientific method, westerners tend to be afraid of feelings and sentiments. It is disconcerting to be told that only by letting go will one know truth … How can the body be a site of knowledge?

Diane Bell learnt from Ngarrindjeri people of 'feeling-as-knowing'. This is knowledge that all humans have always shared, it is just that in the last few hundred years westerners have increasingly felt that they have to justify why feelings are important. I know what I feel about the dormant burrowing frogs waiting in vain for the next cycle of heavy rains and prolonged floods. These frogs sensibly and instinctively dug into the wet clay, with plans to emerge much later after floodwaters penetrate the hardened clay. Burrowing frogs can survive underground for three to five years (Scott 2001, p. 280). Sadly, on the former floodplains the burrowing frogs have dug their own graves. This feeling of loss is knowledge of lives being destroyed as part of the breakdown of connectivity. It is feeling-as-knowing that inspires passion for action.

The importance of moving away from the superiority of rationalist science resonates with scientists who readily admit the subjectivity of their methodologies and the limitations of their knowledge. They see science as a more nuanced approach than the construction of the absolute universal truth (Eagleton 2003, p. 18). As ecologist Frank Egler has observed, 'ecosystems may not only be more complex than we think, they may be more complex than we *can* think' (cited in Rose 2001, pp. 3–4). This is what the precautionary principle is striving at: a principle recommended by scientists to stem the authority of the moderns. Debbie Rose concludes that it is the misrepresentation of scientific knowledge as reductionist truth by governments that has been so effective at empowering destructive modern water habits. The scientists' knowledge that describes water in gigalitres is used by governments to undercut the complexities and connections that scientists are also talking about. The understanding of water in gigalitres kills both the life of water and our capacity to identify our connections with that life (Rose 2007a, pp. 12–13).

The Running the River arrangement is currently informing the river operators how to manage water, including where to store water and which wetlands to 'close down'. If the drought continues, the Murray–Darling Basin Commission has said that managing evaporation along the system will be critical. This evaporation is expected to be three times the minimal volume of water necessary for vital human needs (MDBC 2007a). Because evaporation rates are lower in the deep upstream storages, compared with the extended network of shallow irrigation channels and regulated rivers, river operators are storing water in the Hume and Dartmouth reservoirs. They are also drawing down weir pools such as Lake Mulwala to minimise 'in-river losses' of water to evaporation (MDBC 2007f, p. 3). The operators' decisions about which wetlands to close down or disconnect are also being considered in relation to the regulated flows, weir pool levels, and flow rates of the Murray River System (MDBC 2007d). In 2006, 27 regulated wetlands were closed in South Australia (MDBC 2007a). The more efficient movement, storage and allocation of water within the Murray River System is the outcome of all this work. But it is work that finds it easier to develop more efficient destruction than to address the conceptual deceptions inextricable to the task (Grinde & Johansen 1995, p. 270).

Modern water management has worked in tandem with rationalist market economics, which measures and celebrates economic growth and development but cannot acknowledge ecological destruction. Ecocide is not part of the cost–benefit analysis. Alternatively, the connectivity thinking behind cultural economy takes a critically different approach. Ethical relationships with country are included in the matrix of decision-making. Traditional owners speak about small initiatives, but it is a thinking that can move. Indeed, cultural economy and connectivity are sticky universals: ways of thinking that can be understood by people in their own country.

Ongoing drought has jeopardised The Living Murray's goal of returning a meagre 500 gigalitres of water to the Murray (The Living Murrray 2007). Many water entitlement holders are on zero or very low allocations (MDBC 2007f, p. 5). The death of permanent plantings has already occurred as water shortages affect those farmers with orchards and vineyards (MDBC 2007b, p. 4.). Down in the Murray River town of Mildura these farmers debated mobilising to sue the government for compensation. The government did legislate that it has power over the water, but does this mean it is responsible for the current drought? (Consider McKay 2007, pp. 94–6). In this legal tussle about who is responsible we are being drawn deeper into the world of representations.

Ecological devastation has transformed natural resource management into a moveable feast: policy on the run. The policy-makers keep writing policy as though the humans are in control and the river has no agency, but policy is also changing to acknowledge that there is a problem that needs to be addressed. The politicians are starting to say 'we can't make it rain' and even recommending that we 'pray for rain' (Anon 2007c; Lane 2007). The spectre of global shifts in climate brings an unsettling influence to this now hesitant modern knowledge. Indeed, ongoing drought and record temperatures have already raised the question of whether this drought is in fact a permanent shift in the climate.

This hesitation creates the opportunity to refocus our attention away from the threat of the drought to the threat of our ways of thinking. We need to get better at embracing the complexity, uncertainty and resilience of variable water flows and let go of the false security of regulated flows (Gibbs 2006). We also need to be open to ways of knowing the world that move beyond technical considerations. I argue for humility instead of hubris. We need to peg the authority of modern knowledge at a lower level and be more responsive to the agency of country.

In Australia, we have Indigenous knowledge traditions that elaborate relationships with the river country that know of this agency and communicate with it. Given that we are all living together in the same place, this knowledge is an incredibly valuable perspective for addressing ecological devastation. The traditional owners tell us that instead of perceiving country through modern knowledge we should listen to how country perceives modern knowledge.

ECOLOGICAL DIALOGUE

The cultural flow is not a competition for water. It is a philosophical change in water management that respects a living world within which our lives are embedded in ethical relationships of care. There is no cultural flow from a dead river: on this the ecological philosophers, the traditional owners and the ecologists concur. We must look to our relationships with rivers to understand how to get ourselves out of this catastrophe. The traditional owners do not have 'shared interests' in this work if it kills life. Without a healthy river country there is no point in sitting down at a table with government to discuss fishing rights or moving rocks to repair the fish traps. There is no point going fishing. Without this activity, land use and occupancy mapping becomes an exercise without content.

The theoretical work I have developed in this book encourages us to embrace our complex responses and go beyond representations to bring our focus to our connections with the real world. This involves seeing our own lives as co-dependent and connected with all ecological life. Rather than a static, inert, mute nature that humans can control, manage, engineer and fully understand, we become aware of all the energy and information surrounding and connected to us. Acknowledging all this energy is crucial to addressing the modern misrepresentation of power relationships in the nature/society, human/nonhuman binaries. Timothy Mitchell (2002, p. 53) encourages people who are heavily influenced by modernity to become open to all sorts of agencies and connections by:

> … making this issue of power and agency a question, instead of an answer known in advance. It means acknowledging something of the unresolvable tension, the inseparable mixture, the impossible multiplicity, out of which intention and expertise must emerge.

Modern knowledge has been a powerful influence in Australian natural resource management, but so has country. Country has persistently resisted fitting in with European norms, and scientists and natural resource managers have had to adapt. The variability of Australian ecology has discredited ecological theories about steady state communities and 'natural balance' (Robin 2007, p. 207). The exceptional variability of Australian ecological patterns and connections has challenged ecologists to develop theories that engage with complexity. Extended periods of drought changed federal drought policy to recognise drought as normal. Drought is climate (Botterill 2003).

Bruno Latour theorised how to develop amodern governing principles that check the power of the moderns (Latour 2001, p. 15). Like Timothy Mitchell, Bruno Latour argues that we need to bring the life and agency of all other extra-human actors into the parliament. This is what the traditional owners are saying: that country is part of the dialogue. They express a political and governance system steeped in their lived experiences with country, and enabled and empowered through all these other actors that are ignored by the moderns. Conservation and development projects that cast humans as superior to a subordinate nature, with the power to intervene, are not part of this government.

Many theorists have responded to ecological devastation by stressing the importance of relationships with place (for example, Rose 2000; Main 2005; Robin 2007). The work of place-theorists meets with the ecological philosophers and traditional owners, as all wish to counter objective thinking and to make embodied connections with the world (Malpas 1999, p. 71). Libby Robin writes that we should not get caught up in the rhetoric of sustainability, but engage with how people feel about their local places, because, she argues, sustainability is part of the powerful continuum of scientific knowledge as the method for understanding nature, and is a discussion that is mainly held at national and international scales. Instead, we need a human scale, and not 'human dimensions' that are considered in an abstract way and in bulk. With a human scale, individuals become empowered with moral responsibilities (Robin 2007, p. 217). Monica Morgan is also concerned that responses to ecological devastation continue to be 'quantities and qualities and amounts and money and measurements', but:

> ... the flow of life is not about that ... you can't just measure the breeding cycle of something without understanding what made that fish be there in the first place and what the interrelating components are around it. And that might take them [the scientists] a thousand seasons. [1 July 2004]

Place theory coincides with the scale of knowledge the traditional owners argue for: a return to the perspectives embedded in country. By reconnecting with their local place, people can make connections between their own life and the life of country. This scale empowers people to act and be engaged. It also builds the basis for developing an environmental ethic that is about connections rather than separations. Historian George Main brings place theory to agricultural practices, arguing that we should engage with 'regenerative agriculture' as part of the regeneration of rural place. The term 'regenerative' acknowledges the brutal history of the fragmentation of local ecologies and the current ecological disorder. With this acknowledgment, we can move to prioritise listening and responding to rural places, an activity that requires a deep sensitivity to local ecologies and broad sensual engagement (Main 2005, pp. 245, 249).

Dams built to block the flow of large rivers have been the largest construction projects of the twentieth century, and have had grievous consequences for local people wishing to care for local places. Whilst they were never 'wholly in control', the work of a small number of dam builders has 'altered the distribution of resources across space and time, among entire communities and ecosystems' (Mitchell 2002, pp. 10, 21). The dam builders employed an aggregate thinking that was unresponsive to the diversity of communities and ecosystems living along the rivers. If 'To understand the [Murray] river is to come to terms with Australia' (Larkins & Parish 1982, p. 7), then we must look to the thinking behind past and current management of the Murray, to come to terms with an Australia that has lost its ability to communicate with and celebrate the energy of ecological life.

We need to be open to ecological dialogue. Then, when the river water is so polluted that the river creatures crawl out of it, we can understand that this too is a conversation with country. We can see what is being communicated and respond in a way that sustains the co-dependent relationships. This is an important start for water managers to re-vision relationships with the Murray.

Those of us who are moderns need to listen and respond to life-sustaining connectivities. In doing so, we can understand that there is more than just human culture in the world. Nature has culture too, ways of speaking and expressing regret, dreams, love and reflection. Without the dangerous hubris of our own modern knowledge we could learn to understand these sticky universals.

Down by the river in South Australia, Aboriginal people no longer meet up on Friday afternoons to go fishing or drive around at night in cars with spotlights looking for game. Now when you go down to the river on a nice warm day, nobody is there (Willis et al. 2004, pp. 191, 193). Young people do not know the sociality of river life and the life of country. They have not seen the rivers when they flowed with clear water. It is the Elders who are the witnesses of river devastation who speak up about this. This devastation has happened over their lifetimes. Knowing what they know, they face an urgent task. They are telling us that we need to respect the rhythmic variability of life before it is too late.

I will leave you with the words of Yorta Yorta woman Monica Morgan, who is still concerned that the moderns are not listening to what the traditional owners are saying (Corowa 2006):

> Who else is going to give them the knowledge about protecting country, but those traditional owners who have understood and lived with their country and passed it from generation to generation? Then they're all going to lose, and we're going to lose along with those people.

References

Aboriginal and Torres Strait Islander Commission 2002a, *Offshore water rights discussion*, booklet, Lingiari Foundation, Broome.

—— 2002b, *Onshore water rights discussion*, booklet, Lingiari Foundation, Broome.

Alexandrina Council 2002, 'Sincere expression of sorrow and apology to the Ngarrindjeri people'.

Allen TFH & Hoekstra TW 1992, *Toward a unified ecology*, Columbia University Press, New York.

Altman J 2003a 'People on country, healthy landscapes and sustainable Indigenous economic futures: the Arnhem Land case', *The Drawing Board: An Australian Review of Public Affairs*, vol. 4, no. 2, pp. 65–82.

—— 2003b, 'Promoting Aboriginal economic interests in natural resource management in NSW: Perspectives from tropical north Australia and some prospects', presented at *Relationships between Aboriginal people and land management issues in NSW: barriers and bridges to successful partnerships*, University of Wollongong, 1–3 October 2003.

—— 2004, 'Indigenous interests and water property rights', *Dialogue*, vol. 23, no. 3, pp. 29–33.

—— 2005, 'Development options on Aboriginal lands: Sustainable Indigenous hybrid economies in the twenty-first century', in *The power of knowledge, the resonance of tradition*, L Taylor, GK Ward, G Henderson, R Davis & LA Wallis (eds), Aboriginal Studies Press, Canberra.

Altman J, Buchanan G & Larson L 2007, 'The environmental significance of the Indigenous estate: Natural resource management as economic development in remote Australia', CAEPR, Canberra.

Anderson DG 2000, *Identity and ecology in Arctic Siberia: the Number One Reindeer Brigade*, Oxford University Press, Oxford.

Anderson MK 1997, 'California's endangered peoples and endangered ecosystems', *American Indian Culture and Research Journal*, vol. 21, no. 3, pp. 7–31.

Anon. 2003, 'Worst drought on record', *ABC Rural News*, 3 September 2003.

—— 2005a, 'Towards a culture of nature', Centre for Cross-Cultural Research, Australian National University, 26 July 2005.

—— 2005b, *Daily Advertiser*, 1 July 2005.

REFERENCES

—— 2006, 'Fed: Howard backs draining wetlands to give water to towns', *AAP General News Wire*, 8 November 2006.
—— 2007a, *Canberra Times*, 25 June 2007.
—— 2007b, 'Mighty Murray's majesty', *Sunday Herald*, 12 August 2007, p. 36.
—— 2007c, 'Qld: Beattie says PM's water cuts are needed', *AAP General News Wire*, 19 April 2007
—— n.d., 'The Barmah Millewa Forum', pamphlet, in possession of the author.
Arnold A 2006, 'Turning back the tide of history', *The Sunday Age*, 8 January 2006, p. 18.
Arslan Z 1999, 'Taking rights less seriously: postmodernism and human rights', *Res Publica*, vol. 5, no. 2, pp. 195–215.
Arthur J 2003, *The default country: a lexical cartography of twentieth-century Australia*, UNSW Press, Sydney.
Aslin HJ & Brown VA 2004, *Towards whole of community engagement: a practical toolkit*, MDBC, Canberra.
Atkinson H 2004, 'Yorta Yorta Co-operative Land Management Agreement: impact on the Yorta Yorta Nation', *Indigenous Law Bulletin*, vol. 6, no. 5, pp. 23–5.
ATSIC *see* Aboriginal and Torres Strait Islander Commission.
Attwood B & Foster SG (eds) 2003, *Frontier conflict: the Australian experience*, National Museum of Australia, Canberra.
Australian Academy of Science 1974, *Biological science: the web of life*, Australian Academy of Science, Canberra.
Australian Conservation Foundation 2007, *South Australia: Murray–Darling Wetland report card: the deepening crisis*, Australian Conservation Foundation, Adelaide.
Australian Government Department of Environment and Heritage [2007, <http://wwwdeh.gov.au/esd/national/nsesd/strategy/intro.html>, URL accessed 14 September 2007.
Australian Government 2007, <http://www.nrm.gov.au/nrm/manage.html>, URL accessed 3 September 2007.
Australian Institute of Aboriginal and Torres Strait Islander Studies 2002, *Guidelines for ethical research in Indigenous studies*, Australian Institute of Aboriginal and Torres Strait Islander Studies, Canberra.
Australian Museum 2007, 'Bogong moths fact sheet', <http://www.austmus.gov.au/factsheets/bogong_moths.htm>, URL accessed 24 October 2007.
Baker R, Davies J & Young E (eds) 2001, *Working on country: Contemporary Indigenous management of Australia's lands and coastal regions*, Oxford University Press, Melbourne.
Ball J, Donnelley L, Erlanger P, Evans R, Kollmorgen A, Neal B & M Shirley 2001, 'Inland waters', in *Australia state of the environment report 2001 (theme report)*, CSIRO Publishing on behalf of the Department of Environment and Heritage, Canberra.
Bammer G, Curtis A, Mobbs C, Lane R & Dovers S 2005, 'Australian case studies of integration in natural resource management (NRM)', in *Australasian Journal of Environmental Management: Supplementary Issue*, vol. 12, supplementary issue, pp. 1–64.
Behrendt J & Thompson P 2003, 'The recognition and protection of Aboriginal interests in NSW rivers', *Occasional Paper 1008*, Healthy Rivers Commission of New South Wales, Sydney.
Bell D 1998, *Ngarrindjeri Wurruwarrin: a world that is, was, and will be*, Spinifex Press, Melbourne.

REFERENCES

Bellamy J, Ross H, Ewing S & Meppem T 2002, *Integrated catchment management: learning from the Australian experience for the Murray Darling Basin*, CSIRO Sustainable Ecosystems, Canberra.

Bernstein RJ 1991, *The new constellation: the ethical-political horizons of modernity/postmodernity*, Polity Press, Cambridge.

Birckhead J, De Lacy T & Smith LJ 1993, *Aboriginal involvement in parks and protected areas*, Aboriginal Studies Press, Canberra.

Blackburn K 2002, 'Mapping Aboriginal Nations: the "nation" concept of late nineteenth century anthropologists in Australia', *Aboriginal History*, vol. 26, pp. 131–58.

Blackmore D 2002, 'Protecting the future', in D Connell (ed.), *Uncharted waters*, Murray–Darling Basin Commission, Canberra.

Blackstone W 1979 (1765–69), *Commentaries on the law of England*, vol. II, University of Chicago, Chicago.

Blanch S 1999, 'Environmental flows: present and future', presented at *Australian National Committee on Large Dams* conference, Jindabyne.

Bonta M 2005, 'Becoming-forest, becoming-local: transformations of a protected area in Honduras', *Geoforum*, vol. 36, no. 1, pp. 95–112.

Bonta M & Protevi J 2004, *Deleuze and geophilosophy: a guide and glossary*, Edinburgh University Press, Edinburgh.

Botterill LC 2003, 'Uncertain climate: the recent history of drought policy in Australia', *Australian Journal of Politics and History*, vol. 49, no. 1, pp. 61–74.

Boulton AJ & Brock MA 1999, *Australian freshwater ecology: processes and management*, Gleneagles Publishing, South Australia.

Bradfield S 2004, 'Agreeing to terms: what is a "comprehensive" agreement?', in *Land, rights, laws: issues of native title*, Australian Institute for Aboriginal and Torres Strait Islander Studies, Canberra.

Braun B 2002, *The intemperate rainforest: nature, culture, and power on Canada's west coast*, University of Minnesota Press, Minneapolis and London.

Brennan S, Behrendt L, Strelein L & Williams G 2005, *Treaty*, The Federation Press, Sydney.

Brown W 1995, 'Wounded attachments: late modern oppositional political formations', in J Rajchman (ed.), *The identity in question*, Routledge, New York.

Brown MF 2003, *Who owns native culture?* Harvard University Press, Massachusetts.

Butler J 2003, 'Afterword: after loss, what then?' in DL Eng & D Kazanjian (eds), *Loss: the politics of mourning*, University of California Press, Berkeley.

Carp Control Coordination Group 2000, *National management strategy for carp control 2000–2005*, Murray Darling Basin Commission, Canberra.

Catchment Management Authorities NSW 2004, *Combined NSW catchment management authorities annual report 2003/04*, Catchment Management Authorities NSW, Sydney.

CCCG *see* Carp Control Coordination Group.

Chong J & Ladson AR 2003, 'Analysis and management of unseasonal flooding in the Barmah–Millewa Forest, Australia', *River Research and Applications*, vol. 19, no. 2, pp. 161–80.

Clark I 1995, *Scars in the landscape: a register of massacre sites in western Victoria, 1803–1859*, Aboriginal Studies Press, Canberra.

REFERENCES

Close A 1992, 'The impact of man on the natural flow regime', in N Mackay & D Eastburn (eds), *The Murray*, Murray Darling Basin Commission, Canberra.

CMA *see* Catchment Management Authorities NSW.

COAG *see* Council of Australian Governments.

Commonwealth of Australia 1999, 'National statement on ethical conduct in research involving humans', Canberra.

Connell D 2005a, 'Managing climate for the Murray–Darling Basin (1850–2050)', in T Sherratt, T Griffiths & L Robin (eds), *A change in the weather: climate and culture in Australia*, NMA Press, Canberra.

—— 2005b, 'The chariot wheels of the Commonwealth: the past present and future of inter-jurisdictional water management in the Murray–Darling Basin', PhD thesis, The Australian National University, Canberra.

—— 2007, *Water politics in the Murray–Darling Basin*, The Federation Press, Sydney.

Connell D & Colebatch HK 2006, 'The Murray–Darling Basin – an evolving sphere of public policy', in HK Colebatch (ed.), *Beyond the policy cycle: the policy process in Australia*, Allen & Unwin, Sydney.

Connell, D & Grafton, RQ 2008, 'Planning for water security in the Murray–Darling Basin', *Public Policy*, vol. 3, no. 1, pp. 67–86.

Coorey P 2007, 'For millions the water will stop midyear', *Sydney Morning Herald*, 20 April 2007.

Corowa M 2006, 'Murray, Life + Death', *Message Stick*, ABC TV.

Council of Australian Governments 2004, 'Intergovernmental agreement on a National Water Initiative'.

Crabb P 1997, *Murray–Darling Basin Resources*, Murray–Darling Basin Commission, Canberra.

Craik W 2005, 'Weather, climate, water and sustainable development', *World Meteorological Day Address*, 23 March 2005.

Crenshaw KW 1988, 'Race, reform and retrenchment: transformation and legitimation in anti-discrimination law', *Harvard Law Review*, vol. 101, pp. 1331–87.

Cullen, P 2006, *Science and politics: speaking truth to power*, address to the North American Benthological Society Annual Conference, June 2006.

DAA *see* Department of Aboriginal Affairs.

DEH, *see* Australian Government Department of Environment and Heritage.

DEWHA, *see* Department of Environment, Water, Heritage and the Arts.

Department for Environment and Heritage 2000, *Coorong, and Lakes Alexandrina and Albert Ramsar Management Plan*, South Australian Department for Environment and Heritage, Adelaide.

Department of Aboriginal Affairs 2004, *NSW Aboriginal people acquiring and managing land for conservation purposes*, NSW Department of Aboriginal Affairs, Sydney.

Department of Infrastructure Planning and Natural Resources 2004, *Water sharing plan for the Murrumbidgee regulated river water source*, Department of Infrastructure, Planning and Natural Resources.

Department of Water, Land and Biodiversity Conservation n.d., *Murray Mouth sand pumping project*, Fact Sheet 23, Government of South Australia.

DIPNR *see* Department of Infrastructure Planning and Natural Resources.

Dodson M 2003, 'The end in the beginning: re(de)finding Aboriginality', in M Grossman (ed.), *Blacklines: contemporary critical writing by Indigenous Australians*, Melbourne University Press, Melbourne.

Dodson M & Strelein L 2001, 'Australia's nation building: renegotiating the relationship between indigenous peoples and the state', *UNSW Law Journal*, vol. 24, no. 3, pp. 826–39.

Dovers S 2003, 'Discrete, consultative policy processes: lessons from the National Conservation Strategy for Australia and National Strategy for Ecologically Sustainable Development', in S Dovers, & S Wild River (eds), *Managing Australia's environment*, The Federation Press, Sydney.

DWLBC *see* Department of Water, Land and Biodiversity Conservation.

Dyer F, Thoms M & Norris R 2003, 'Can environmental flows compensate for upstream dams?', *Watershed*, February, pp. 4–7.

Eagleton T 2003, *After theory*, Allen Lane, London.

Eastburn D 1992, 'The river', in N Mackay & D Eastburn (eds), *The Murray*, Murray–Darling Basin Commission, Canberra.

Department of Environment, Water, Heritage and the Arts, 1992, *National strategy for ecologically sustainable development*, prepared by the Ecologically Sustainable Development Steering Committee, endorsed by the Council of Australian Governments.

Elder B 1988, *Blood on the wattle: massacres and maltreatment of Australian Aborigines since 1788*, Child & Associates, Sydney.

Elkin A 1970, 'Before it is too late', in R Berndt (ed.), *Australian Aboriginal Anthropology*, University of Western Australia Press, Nedlands.

Eng, DL & Kazanjian, D (eds) 2003, *Loss: the politics of mourning*, University of California Press, Berkeley.

English A 2002, 'More than archaeology: developing comprehensive approaches to Aboriginal heritage management in NSW', *Australian Journal of Environmental Management*, vol. 9, no. 4, pp. 218–27.

Eriksen TH 2001, 'Between universalism and relativism: a critique of the UNESCO concept of culture', in JK Cowan, M-B Dembour & RA Wilson (eds), *Culture and rights: anthropological perspectives*, Cambridge University Press, Cambridge, pp. 127–48.

Fabian J 1991, *Time and the work of anthropology: critical essays 1971–1991*, Harwood Academic Publishers, Chur, Switzerland.

Farley Consulting Group 2003, *Indigenous response to The Living Murray initiative*, Murray Darling Basin Commission, Canberra.

FCG *see* Farley Consulting Group.

Flood J 1976, 'Man and ecology in the highlands of southeastern Australia: a case study', in N Peterson (ed.), *Tribes and boundaries in Australia*, Australian Institute of Aboriginal Studies, Canberra.

Forward NRM, Arrilla-Aboriginal Training & Development 2003, 'Scoping study on Indigenous involvement in natural resource management decision-making and the integration of Indigenous cultural heritage considerations into relevant Murray–Darling Basin Commission programs', Murray–Darling Basin Commission, Canberra.

REFERENCES

Gammage B 1986, *Narrandera Shire*, Bill Gammage for the Narrandera Shire Council.

Gehrke P, Gawne B & Cullen P 2003, *What is the status of river health in the Murray–Darling Basin?*, CSIRO Land and Water.

Gell P 2007, 'River Murray wetlands: past and future' in E Potter, S McKenzie, A Mackinnon & J McKay (eds), *Fresh water: new perspectives on water in Australia*, Melbourne University Press, Melbourne.

Gibbs L 2006, 'Valuing water: variability and the Lake Eyre Basin, central Australia', *Australian Geographer*, vol. 37, no. 1, pp. 73–83.

Gilmore M 1963 (1934), *Old days: old ways — a book of recollections*, Angus & Robertson, Sydney.

Goodall H 1996, *Invasion to Embassy: land in Aboriginal politics in New South Wales, 1770–1972*, Allen & Unwin (in association with Black Books), Sydney.

Green R 2002, 'Preface', in D Connell (ed.), *Uncharted waters*, Murray–Darling Basin Commission, Canberra.

Griffiths T 1997, 'Ecology and empire: towards an Australian history of the world', in T Griffiths & L Robin (eds), *Ecology and empire: environmental history of settler societies*, Melbourne University Press, Melbourne.

Grinde D & Johansen B 1995, *Ecocide of Native America: environmental destruction of Indian lands and peoples*, Clear Light Publishers, Santa Fe.

Haila Y 2000, 'Beyond the nature–culture dualism', *Biology and Philosophy*, no. 15, pp. 155–75.

Hancock D 2003, 'Top End's fragile balance on the brink', *The Age*, 15 September 2003, p. 8.

Haraway D 1988, 'Situated knowledges: the science question in feminism and the privilege of partial perspective, *Feminist Studies*, v. 14, no. 3, pp. 575–99.

—— 1992, 'The promises of monsters: a regenerative politics for inappropriate/d others', in C Nelson & PA Treichler (eds), *Cultural Studies*, Routledge, New York.

Harman C & Stewardson M 2005, 'Optimizing dam release rules to meet environmental flow targets', *River Research and Applications*, no. 21, pp. 113–29.

Harmsen N 2007, 'Things are not looking good for River Murray', ABC News, 23 August 2007.

Harries-Jones P 1995, *A recursive vision: ecological understanding and Gregory Bateson*. University of Toronto Press, Toronto.

Hattam R, Rigney D & Hemming S 2007, 'Reconciliation? Culture, nature and the Murray River', in E Potter, S McKenzie, A Mackinnon & J McKay (eds), *Fresh water: new perspectives on water in Australia*, Melbourne University Press, Melbourne.

Hemming S 2005, 'Managing cultures into the past', presented at *Understanding cultural landscapes symposium*, Humanities Research Centre for Cultural Heritage and Cultural Exchange, Flinders University.

Hemming S, Rigney D & Pearce M 2007, 'Justice, culture and economy for the Ngarrindjeri nation', in E Potter, S McKenzie, A Mackinnon & J McKay (eds), *Fresh water: new perspectives on water in Australia*, Melbourne University Press, Melbourne.

Hennessy K, McInnes K, Abbs D, Jones R, Bathols J, Suppiah R, Ricketts J, Rafter T, Collins D & Jones D 2004, *Climate change in New South Wales, Part 2: Projected changes in climate extremes*, CSIRO.

Hinkson M & Smith BR 2005, 'Introduction: conceptual moves towards an intercultural analysis', *Oceania*, vol. 75, no. 3, pp. 157–66.
Hoffmeyer J 1997, *Signs of meaning in the universe*, Indiana University Press, Bloomington.
Hone P, Fraser I & Haszler H 2001, *Land retirement as a conservation policy*, Land and Water Resources Research and Development Corporation.
Howard J 2007, *A national plan for water security*, 25 January 2007.
Howitt R 2001, *Rethinking resource management: justice, sustainability, and indigenous peoples*, Routledge, London.
Howitt R & Suchet-Pearson S 2006, 'Rethinking the building blocks: ontological pluralism and the idea of "management"', *Geography Annual*, vol. 88B, no. 3, pp. 323–35.
Humphries P, King AJ & Koehn JD 1999, 'Fish, flows and floodplains: links between freshwater fishes and their environment in the Murray–Darling River system, Australia', *Environmental Biology of Fishes*, vol. 56, pp. 129–51.
Hunn ES, Johnson DR, Russell PN & Thornton TF 2003, 'Huna Tlingit traditional environmental knowledge, conservation, and the management of a "wilderness" park', *Current Anthropology*, vol. 44, supplement, S79–103.
Ignatieff M 2003, *Human rights as politics and idolatry*, Princeton University Press, Princeton.
Ingold T (ed.) 1996a, *Key debates in anthropology*, Routledge, London.
—— 1996b, 'Part II: the debate', in T Ingold (ed.), *Key debates in anthropology*, Routledge, London.
—— 2000, *The perception of the environment: essays in livelihood, dwelling and skill*, Routledge, London.
Jackson S 2005, 'Indigenous values and water resource management: a case study from the Northern Territory', *Australasian Journal of Environmental Management*, vol. 12, no. 3, pp. 136–46.
Jackson S & Morrison J 2007, 'Indigenous perspectives in water management, reforms and implementation', in K Hussey & S Dovers (eds), *Managing water for Australia: the social and institutional challenges*, CSIRO Publishing, Melbourne.
Jenkin C 2007a, 'Cheaper to steal water and pay fine', *The Advertiser*, 26 September 2007, p. 3.
—— 2007b, 'Houseboat hardship: River Murray', *The Advertiser*, 22 September 2007, p. 56.
Jones G, Hillman TJ, Kingsford R, McMahon T, Walker K, Arthington AH, Whittington J & Cartwright, S 2002, 'Independent report of the expert reference panel on environmental flows and water quality requirements for the River Murray System', *Report to the Murray–Darling Basin Ministerial Council*, Cooperative Research Centre for Freshwater Ecology.
Jones H 2002, 'Personal statement', in D Connell (ed.), *Uncharted waters*, Murray–Darling Basin Commission, Canberra.
Kalland A 2003, 'Anthropology and the concept "sustainability": some reflections', in A Roepstorff, N Bubandt & K Kull (eds), *Imagining nature: practices of cosmology and identity*, Aarhus University Press, Aarhus.
Kayano S 1994 (1980), *Our land was a forest* Westview Press, Boulder.
Keen I 2006, *Aboriginal economy and society: Australia at the threshold of colonisation*, Oxford University Press, Melbourne.

REFERENCES

Kingsford R 2000, 'Ecological impacts of dams, water diversions and river management on floodplain wetlands in Australia', *Austral Ecology*, vol. 25, pp. 109–27.

—— 2002, *Inland rivers and floodplains*, Land & Water Australia, Canberra.

Kingsford RT 2003, 'Ecological impacts and institutional and economic drivers for water resource development: a case study of the Murrumbidgee River, Australia' *Ecosystem Health & Management*, vol. 6, no. 1, pp. 69–79.

Kinnane S 2002, 'Recurring visions of Australindia', in A Gaynor, M Trinca & A Haebich (eds), *Country: visions of land and people in Western Australia*, Western Australian Museum, Perth.

Lane S 2007, 'PM urges nation to pray for rain', *Australian Broadcasting Corporation*, 16 May 2007.

Langton M 1995, 'Arts, wilderness and terra nullius', in R Sultan (ed.), *Ecopolitics IX: perspectives on Indigenous people's management of environment resources*, Northern Land Council, Darwin.

—— 2002, 'The edge of the sacred, the edge of death: sensual inscriptions', in B David & W Meredith (eds), *Inscribed landscapes: marking and making place*, University of Hawaii Press, Hawaii.

Langton M, Mazel O, Palmer L, Shain K & Tehan M 2006 (eds), *Settling with Indigenous peoples*, The Federation Press, Sydney.

Larkins J & Parish S 1982, *Australia's greatest river: and the Murray from source to sea*, Rigby Publishers, Adelaide.

Latour B 2001 (1993), *We have never been modern*, Harvard University Press, Cambridge.

Leopold A 1949, *A Sand County almanac*, Oxford University Press, New York.

Leslie DJ 2001, 'Effects of river management on colonially-nesting waterbirds in the Barmah–Millewa forest, south-eastern Australia', *Regulated Rivers: Research & Management*, vol. 17, no. 1, pp. 21–36.

Lewis D 2006, 'Fat ducks, fat cattle — fat chance', *Sydney Morning Herald*, 8–9 July 2006, p. 31.

Lindenmayer DB 2007, *On borrowed time: Australia's environmental crisis and what we must do about it*, Penguin Books (in association with CSIRO Publishing), Camberwell.

Lingiari Foundation (ed.) 2002, *Background briefing papers: Indigenous rights to water*, Lingiari Foundation, Broome.

Littlewood R 1996, 'For the motion', in T Ingold (ed.), *Key debates in anthropology*, Routledge, London.

Lohmann L 1993, *Green Orientalism*, The Corner House, Dorset, UK.

Lourandos H 1987, 'Swamp managers of southwestern Victoria', in DJ Mulvaney (ed.), *Australians to 1788*, Fairfax, Syme & Weldon, Sydney.

Macklem P 1991, 'First Nations self-government and the borders of the Canadian legal imagination', *McGill Law Journal*, vol. 36, no. 2, pp. 382–456.

Main G 2005, *Heartland: the regeneration of rural place*, UNSW Press, Sydney.

Malpas JE 1999, *Place and experience: a philosophical topography*, Cambridge University Press, Cambridge.

Manning AD, Lindenmayer DB & Nix HA 2004, 'Continua and Umwelt: novel perspectives on viewing landscapes', *Oikos*, vol. 104, no. 3, pp. 621–28.

Marcus GE 1995, 'Ethnography in/of the world system: the emergence of multi-sited ethnography', *Annual Review of Anthropology*, vol. 1995, no. 24, pp. 95–117.

Marcus J 2004, 'Book review: "The Heritage of Hindmarsh Island"', *The Australian Journal of Anthropology*, vol. 15, no. 3, pp. 331–43.

Mares P 2007, 'The unmentionable water option', *National Interest*, ABC Radio National, 5 August 2005.

Martin D 2003, 'Rethinking the design of indigenous organisations: the need for strategic engagement', in *CAEPR Discussion Paper 248*, Australian National University, Canberra.

Mathews F 1994, *The ecological self*, Routledge, London.

McKay J 2002, *Encountering the South Australian landscape: early European misconceptions and our present water problems*, Hawke Institute, Magill.

—— 2006–07, 'The quest for environmentally sustainable water use: constitutional issues for federal, state and local government', *Reform*, vol. 89, Summer 2007, pp. 22–7.

—— 2007, 'Water, rivers and ecologically sustainable development', in E Potter, S McKenzie, A Mackinnon & J McKay (eds), *Fresh water: new perspectives on water in Australia*, Melbourne University Press, Melbourne.

McMullen J 2007, 'The lie of the land', *Difference of Opinion*, ABC TV, 4 October 2007.

MDBC *see* Murray–Darling Basin Commission.

MDBMC *see* Murray–Darling Basin Ministerial Council.

Merlan F 2005, 'Explorations towards intercultural accounts of socio-cultural reproduction and change', *Oceania*, vol. 75, no. 3, pp. 167–82.

Mitchell T 2000, 'The stage of modernity', in T Mitchell (ed.), *Questions of modernity*, University of Minnesota Press, Minneapolis and London.

—— 2002, *Rule of experts: Egypt, techno-politics, modernity*, University of California Press, Berkeley and Los Angeles.

MLDRIN *see* Murray Lower Darling Rivers Indigenous Nations.

Morgan M, Strelein L & Weir J 2004, 'Indigenous rights to water in the Murray–Darling Basin: in support of the Indigenous final report to The Living Murray initiative', in *Research Discussion Paper No.14*, AIATSIS, Canberra.

Morgan M, Strelein L & Weir J 2006, 'Authority, knowledge and values: Indigenous Nations engagement in the management of natural resources in the Murray–Darling Basin', in M Langton, O Mazel, L Palmer, K Shain & M Tehan (eds), *Settling with Indigenous peoples*, The Federation Press, Sydney.

Morphy H 1995, 'Landscape and the reproduction of the ancestral past', in E Hirsch & M O'Hanlon (eds), *Between place and space*, Oxford University Press, Oxford.

Muecke S 2004, *Ancient & modern: time, culture and indigenous philosophy*, UNSW Press, Sydney.

Muir J & Morgan M 2002, 'Yorta Yorta: the community's perspective on the treatment of oral history', in M Paul & G Gray (eds), *Through a smoky mirror: history and native title*, Aboriginal Studies Press, Canberra.

Murdoch J 1997, 'Inhuman/nonhuman/human: actor-network theory and the prospects for a non-dualistic and symmetrical perspective on nature and society', *Environment and Planning D: Society and Space*, vol. 15, no. 6, pp. 731–56.

REFERENCES

Murray–Darling Basin Commission 2000, *Review of the operation of the cap: overview report of the Murray–Darling Basin Commission*, Murray–Darling Basin Commission, Canberra.

—— 2005, *The Living Murray foundation report on the significant ecological assets targeted in the First Step Decision*, Murray–Darling Basin Commission, Canberra.

—— 2006, *River Murray System: drought update*.

—— 2007a, 'Fact sheet: drought contingency measures', May 2007.

—— 2007b, 'Dry inflow contingency planning', *Overview report to first ministers*, September 2007.

—— 2007c, *First phase of water purchase pilot closes early because of strong response*, press release, 13 August 2007.

—— 2007d, *New Murray wetlands atlas helps map crucial drought water flows*, press release, 11 October 2007.

—— 2007e, *River Murray System: drought update no. 8*, June 2007.

—— 2007f, *River Murray System: drought update no. 10*, October 2007.

—— n.d., *Indigenous Basin-wide gathering*, conference papers, presentations and outcomes, <http://www.mldrin.org.au/about/default.htm>, URL accessed 14 September 2007.

—— 2007g, <http://www.mdbc.gov.au/river_murray/river_murray_system/river_murray_system.htm>, URL accessed 6 July 2007. This website is now archived.

—— 2007h, <http://www.mdbc.gov.au/rmw/running_the_river>, URL accessed 6 July 2007. This website is now archived.

—— 2007i, <http://thelivingmurray.mdbc.gov.au/programs/environmental_works_and_measures>, URL accessed 26 October 2007.

Murray–Darling Basin Ministerial Council 1987, *Murray–Darling Basin environmental resources study*, State Pollution Control Commission, Sydney.

—— 2001, *Integrated catchment management in the Murray–Darling Basin 2001–2010: delivering a sustainable future*, Murray–Darling Basin Ministerial Council, Canberra.

—— 2002, *The Living Murray: a discussion paper on restoring the health of the River Murray, Stage 1: Informing and engaging the community*, Murray–Darling Basin Commission, Canberra.

—— 2003a, 'Murray–Darling Basin Ministerial Council Communiqué', 14 November 2003.

—— 2003b, *Native fish strategy for the Murray–Darling Basin 2003–2013*, Murray–Darling Basin Commission, Canberra.

—— 2005a, 'Communiqué', 17 October 2005.

—— 2005b, *The Living Murray business plan*, Murray–Darling Basin Commission, Canberra.

Murray Lower Darling Rivers Indigenous Nations n.d., 'Introduction to the Murray Lower Darling Rivers Indigenous Peoples', in possession of the author.

—— 2007a, <http://www.mldrin.org.au/whatwedo/wayforward.htm>, URL accessed 1 October 2007.

—— 2007b, 'A cooperative agreement between Murray Lower Darling Rivers Indigenous Nations and Environmental Non-Government Organisations', 23 February 2007.

—— 2009, <http://www.mldrin.org.au/about/aboutlogo.htm>, URL accessed 12 May 2008.

National Parks Association of NSW 2007, <http://www.npansw.org.au/web/conservation/western/redgum/index.htm>, URL accessed 13 September 2007.

National Waterexchange 2007, <https://www.waterexchange.com.au>, URL accessed 26 October 2007.

NCCMA *see* North Central Catchment Management Authority.

North Central Catchment Management Authority, North West Nations Clans Aboriginal Corporation & Yorta Yorta Nation Aboriginal Corporation 2002, *Protocols, principles and strategies agreement for Indigenous involvement in land and water management*, The State of Victoria, Melbourne.

NPA *see* National Parks Association of NSW.

Ogle G 2002, 'Just when you thought it was safe to talk about Hindmarsh Island', *Indigenous Law Bulletin*, vol. 5, no. 15, pp. 16–18..

Palmer L 2006, ' "Nature", place, and the recognition of Indigenous polities', *Australian Geographer*, vol. 37, no. 1, pp. 33–43.

Pappin, M 2004, 'Up the river forum', *Message Sticks Festival*, Sydney Opera House, May 2004.

Pascoe B 2007, *Convincing ground: learning to fall in love with your country*, Aboriginal Studies Press, Canberra.

Paz O 1967, *The labyrinth of solitude*, Penguin, London.

Plumwood V 2002 (1993), *Feminism and the mastery of nature*, Routledge, London & New York.

Povinelli EA 1993, *Labor's lot: the power, history, and culture of Aboriginal action*, University of Chicago Press, Chicago.

—— 2002, *The cunning of recognition: indigenous alterities and the making of Australian multiculturalism*, Duke University Press, Durham & London.

Powell JM 1989 *Watering the Garden State: water, land and community in Victoria, 1834–1988*, Allen & Unwin, Sydney.

Productivity Commission 2003, 'Water rights arrangements in Australia and overseas', in *Commission Research Paper*, Productivity Commission, Melbourne.

—— 2006, 'Rural water use and the environment: the role of market mechanisms', *Research Report*, Productivity Commission, Melbourne.

Proust K 2003, 'Ignoring the signals: irrigation salinity in New South Wales, Australia', *Irrigation and Drainage*, vol. 52, no. 2003, pp. 39–49.

Pyle RM 1992, 'Intimate relations and the extinction of experience', *Left Bank*, vol. 2, pp. 61–9.

Read P 1988, *A hundred years war: the Wiradjuri people and the state*, Australian National University Press, Sydney.

—— 1996, *Returning to nothing: the meaning of lost places*, Cambridge University Press, Melbourne.

Reid MA & Brooks JJ 2000, 'Detecting effects of environmental water allocations in wetlands of the Murray–Darling Basin, Australia', *Regulated rivers: research & management*, vol. 16, no. 2000, pp. 479–96.

REFERENCES

Ridgeway, A 2005, 'Addressing the economic exclusion of Indigenous Australians through native title', *Mabo Lecture, National Native Title Conference*, Coffs Harbour, 2005, <http://ntru.aiatsis.gov.au/conf2005/papers/papers.html>, URL accessed 9 September 2007.

Rigby, K 2004, *Topographies of the sacred: the poetics of place in European romanticism*, University of Virginia Press, Charlottesville.

Ritter D 2004, 'The judgement of the world: the Yorta Yorta case and the "tide of history"', *Australian Historical Studies*, no. 123, pp. 106–21.

Robin L 2007, *How a continent created a nation*, UNSW Press, Sydney.

Rose DB 1996, *Nourishing terrains: Australian Aboriginal views of landscape and wilderness*, Australian Heritage Commission, Canberra.

—— 1999, 'Indigenous ecologies and an ethic of connection', in N Low (ed.), *Global ethics and environment*, Routledge, London.

—— 2000, 'Writing place', in A Curthoys & A McGrath (eds), *Writing histories: imagination and narration*, Monash Publications in History, School of Historical Studies, Monash University, Clayton, Victoria.

—— 2001, 'The silence and power of women', in P Brock (ed.s), *Aboriginal women, politics and land*, Allen & Unwin, Sydney.

—— 2004a, 'Fresh water rights and biophillia: Indigenous Australian perspectives', *Dialogue*, vol. 23, no. 3, pp. 35–43.

—— 2004b, *Reports from a wild country: ethics for decolonisation*, UNSW Press, Sydney.

—— 2007a, 'Justice and longing', in E Potter, S McKenzie, A Mackinnon & J McKay (eds), *Fresh water: new perspectives on water in Australia*, Melbourne University Press, Melbourne.

—— 2007b, 'Connecting nature and culture: the role of the humanities', presentation as part of the *Fenner School of Environment and Society* seminar series, Australian National University, Canberra, 31 May 2007.

Rose DB, James D & Watson C 2003, *Indigenous kinship with the natural world in New South Wales*, NSW National Parks and Wildlife Service, Sydney.

Ross S 2006–07, 'Caring for country: Indigenous nations and water management', *Reform*, vol. 89, Summer 2007, pp. 18–21.

Rowse T 2002, *Indigenous futures: choice and development for Aboriginal and Islander Australia*, UNSW Press, Sydney.

Sahlins M 1999, 'What is anthropological enlightenment? Some lessons of the twentieth century', *Annual Review of Anthropology*, vol. 28, pp. i–xxiii.

Scott AC 2001, 'Other riverine animals', in WJ Young (ed.), *Rivers as ecological systems: the Murray–Darling Basin*, Murray–Darling Basin Commission, Canberra.

Scott J 1998, *Seeing like a state: how certain schemes to improve the human condition have failed*, Yale University Press, Newhaven & London.

Sillitoe P 2007, 'Local science vs. global science: an overview', in P Sillitoe (ed.), *Local science vs. global science: approaches to indigenous knowledge in international development*, Berghahn Books, New York.

Sinclair P 2001, *The Murray: a river and its people*, Melbourne University Press, Melbourne.

Smith BR 2000, 'Local and "diaspora" connections to country and kin in central Cape York Peninsula', in *Land, rights, laws: issues of native title*, Australian Institute for Aboriginal and Torres Strait Islander Studies, Canberra.

—— 2003, 'Whither "certainty"? Coexistence, change and land rights in northern Queensland, *Anthropological Forum*, vol. 13, no. 1, pp. 27–48.

—— 2005, ' "We got our own management": local knowledge, government and development in Cape York Peninsula', *Australian Aboriginal Studies*, vol. 2005, no. 2, pp. 4–15.

—— 2006, ' "More than love": Locality and affects of indigeneity in northern Queensland', *The Asia Pacific Journal of Anthropology*, vol. 7, no. 3, pp. 221–35.

—— 2007, ' "Indigenous" and "scientific" knowledge in central Cape York Peninsula', in P Sillitoe (ed.), *Local science vs. global science: approaches to indigenous knowledge in international development*, Berghahn Books, New York.

—— forthcoming, 'Managing in a new environment: ecologies and resources in post-colonial north Queensland', *Cultural Studies Review*, vol. 12, no. 1.

Smith DI 2001, *Water in Australia: resources and management*, Oxford University Press, Melbourne.

Stanner WEH (ed.) 1989 (1963), *On Aboriginal religion*, University of Sydney, Sydney.

Strang V 2006, *The meaning of water*, Berg, Oxford & New York.

Strathern M 1980, 'No nature, no culture: the Hagan case', in CP MacCormack & M Strathern (eds), *Nature, culture and gender*, Cambridge University Press, Cambridge.

Strelein L, Dodson M & Weir J 2001, 'Understanding non-discrimination: native title law and policy in a human rights context', *Balayi: Culture, Law and Colonialism*, vol. 3, pp. 113–48.

Strelein L 1998, 'Indigenous self-determination claims and the common law in Australia', PhD thesis, Australian National University, Canberra.

—— 2000, 'The "courts of the conqueror": the judicial system and the assertion of Indigenous people's rights', *Australian Indigenous Law Reporter*, vol. 5, no. 3, pp. 1–23.

—— 2001, 'Conceptualising native title', *The Sydney Law Review*, vol. 23, no. 1, pp. 95–124.

—— 2002–03, 'Missed meanings: the language of sovereignty in the treaty debate', *Arena*, vol. 20, pp. 83–96

—— 2005, 'Culture and commerce: the use of fishing traditions to prove native title', in L Taylor, GK Ward, G Henderson, R Davis & LA Wallis (eds), *The power of knowledge, the resonance of tradition*, Aboriginal Studies Press, Canberra.

—— 2006, *Compromised jurisprudence: native title cases since* Mabo, Aboriginal Studies Press, Canberra.

—— 2007, 'Being sovereign — holding native title: examining the aspirations for prescribed body corporates', presentation as part of the *Centre for Aboriginal Economic Policy Research* seminar series, Australian National University, Canberra, 3 October 2007.

Sutton P 1995, *Country: Aboriginal boundaries and land ownership in Australia*. Aboriginal History Monograph, v. 1995, no. 3, The Australian National University, Canberra.

Taylor J & Biddle N 2004, *Indigenous people in the Murray–Darling Basin: a statistical profile*, CAEPR, Canberra.

The Living Murray 2007, 'Progress report', <http://thelivingmurray.mdbc.gov.au/programs/water_recovery/progress>, URL accessed 13 September 2007.

REFERENCES

Thoms M, Maher S, Terrill P, Crabb P, Harris J & Sheldon F 2004, 'Environmental flows in the Darling River', in R Breckwoldt, R Boden & J Andrew (eds), *The Darling*, Murray–Darling Basin Commission, Canberra.

Tischendorf L & Fahrig L 2000, 'On the usage and measurement of landscape connectivity', *Oikos*, no. 90, pp. 7–19.

Tobias T 2000, *Chief Kerry's moose: a guidebook to land use and occupancy mapping, research design and data collection*, Union of BC Indian Chiefs and Ecotrust, Canada.

Toussaint S 2006, 'Introducing water: a symposium, and this volume', in M Leybourne & A Gaynor (eds), *Water: histories, cultures, ecologies*, University of Western Australia Press, Perth.

Tsing A 2005, *Friction: an ethnography of global connection*, Princeton University Press, Princeton & Oxford.

Tully J 2004a (1995), *Strange multiplicity: constitutionalism in an age of diversity*, Cambridge University Press, Cambridge & New York.

—— 2004b, 'Recognition and dialogue: the emergence of a new field', *Critical Review of International Social and Political Philosophy*, no. 7, pp. 84–106.

Turnbull M 2007, *$7 million boost for Indigenous environmental projects*, Minister for the Environment and Water Resources, media release.

UNCESCR *see* United Nations Committee on Economic Social and Cultural Rights.

United Nations Committee on Economic Social and Cultural Rights 2002, *General comment no.15*, United Nations Economic and Social Council.

VEAC *see* Victorian Environmental Assessment Council.

Victorian Environmental Assessment Council 2007, *River red gum forests investigation: draft proposals paper for public comment*, Victorian Environmental Assessment Council, Melbourne.

Wahlquist A 2005, 'Flood spawns breeding of Murray's native perch', *The Australian*, 23 November 2005, p. 5.

—— 2007, 'Price leaps as drought effects bite', *Weekend Australian*, 13 October 2007, p. 1.

Ward JV, Tockner K & Schiemer F 1999, 'Biodiversity of floodplain river ecosystems: ecotones and connectivity', *Regulated Rivers: Research & Management*, vol. 15, no. 1–3, pp. 125–39.

Ward, N in press, 'Good methodology travels: an Australian case study, in TN Tobias 2009, *Living proof: the essential data-collection guide for Indigenous use-and-occupancy map surveys*, The Union of BC Indian Chiefs and Ecotrust Canada, Vancouver, British Columbia.

Warshall P 2002, 'Watershed governance', in D Rothernberg & M Ulvaeus (eds), *Writing on water*, MIT Press, Massachusetts.

Webb S 1984, 'Intensification, population and social change in southeastern Australia: the skeletal evidence', *Aboriginal History*, vol. 8, no. 2, pp. 154–72.

Weir J 2006, 'Cultural flows in the Murray Lower Darling rivers', in S Jackson (ed.), *Recognising and protecting Indigenous values in water resource management*, CSIRO, Darwin.

—— 2007a, 'Native title and governance: the emerging corporate sector prescribed for native title holders', *Land, rights, laws: issues of native title*, Australian Institute for Aboriginal and Torres Strait Islander Studies, Canberra.

—— 2007b, 'The traditional owner experience along the Murray River', in E Potter, S McKenzie, A Mackinnon & J McKay (eds), *Fresh water: new perspectives on water in Australia*, Melbourne University Press, Melbourne.

Weir J & Ross S 2007, 'Murray Lower Darling Rivers Indigenous Nations', in F Morphy & BR Smith (eds), *The effects of native title*, ANU E Press, Canberra.

Williams RA 1990, 'Encounters on the frontiers of international human rights law: redefining the terms of indigenous people's survival in the world', *Duke Law Journal*, no. 1990, pp. 660–704.

Willis E, Pearce M & Jenkin T 2004, 'The demise of the Murray River: insights into lifestyle, health and well-being for rural Aboriginal people in the Riverland', *Health Sociology Review*, vol. 13, pp. 187–97.

Yencken D & Connell D 2002, 'A winding road: an assessment of progress towards sustainability in the Murray–Darling Basin', in D Connell (ed.), *Uncharted waters*, Murray–Darling Basin Commission, Canberra.

Young WJ 2001, 'Algae, bacteria and fungi', in WJ Young (ed.), *Rivers as ecological systems: the Murray–Darling Basin*, Murray–Darling Basin Commission, Canberra.

Young WJ & Hillman TJ 2001, 'A tale of two rivers', in WJ Young (ed.), *Rivers as ecological systems: the Murray–Darling Basin*, Murray–Darling Basin Commission, Canberra.

Young WJ, Schiller CB, Harris JH, Roberts J & Hillman TJ. 2001a, 'River flow, processes, habitats and river life', in WJ Young (ed.), *Rivers as ecological systems: the Murray–Darling Basin*, Murray–Darling Basin Commission, Canberra.

Young WJ, Schiller CB, Roberts J & Hillman TJ 2001b, 'The rivers of the basin and how they work', in WJ Young (ed.), *Rivers as ecological systems: the Murray–Darling Basin*, Murray–Darling Basin Commission, Canberra.

Index

(numbers in italics refer to images)

Aboriginal and Torres Strait Islander Commission, 87
Aboriginal identity, *see* identity
Aboriginal knowledge, *see* Indigenous knowledge
Aboriginal sites, *see* places
acknowledgement of ecocide, 134–41
agriculture, 6–10, 31–46, 140, 144
   in Commission's meaning of 'natural resource management', 72
   irrigators' description of Barmah–Millewa wetlands, 127–8
   in national identity narratives, 3
   Northern Territory, 119
   Olney's argument against Yorta Yorta adopting commercial practices, 74–5
   'regenerative', 147
   settlers, 3–4, 6–7, 31–2
   see also forests and forestry; infrastructure and water storage; water trading
Alexandrina Council, 85
algae, blue-green, 36, 60, 120, 127, 136
allocations, *see* water entitlements
Altman, Jon, 131–2, 133
amodern knowledge, xii, 115, 126
   Latour's arguments, 19, 146
   see also connectivity; Indigenous knowledge
ancestors and ancestral beings, 11–14
   see also Dreaming stories
Anderson, David, 50
Anderson, M Kat, 58–9
animals, 13, 54, 140
   adaptation to variability, 29
   anthrozoology, 50
   Bogong moths, 63, 140
   colonialists' perceptions, 6

   environmental flows and, 126
   frogs, 140, 143
   land use and occupancy mapping, 2, 107–8
   in modern thinking, 4
   in older forms of English common law, 45
   ramifications of extinction in terms of identity, 58–9
   totemic, 121
   see also birds; fish; human beings
anthropology, 18, 20–1, 55
anthrozoology, 50
aquifers (groundwater), 33, 39, 43, 52
artistic responses to water management challenges, 18–19
Aswan Dam, 5, 17
Atkinson, Henry, 82, 85, 89–90, 99
   address to MLDRIN meeting in Barmah forest, 98
   on Barmah–Millewa Forum, 68
   on degradation of country, 60, 61
   on meeting leading to creation of MLDRIN, 92
   reminiscences, 51
   on water flows, 120, 122–3; environmental, 127
   on Yorta Yorta native title decision, 75
Australian Conservation Foundation, 101
Australian Constitution, 33
Australian identity, *see* national identity

Barapa Barapa Nation, 93, 94
Barkandji Nation, 97–8
   Mitchell, Junette, 133, 136
Barmah–Millewa, 30, 35, 36–7, plate [5] between 112–13, 127–8
   Henry Atkinson's descriptions, 51, 60

INDEX

Lee Joachim's description of lake, 13
Land Management Agreement with Victorian Government, 85
land use and occupancy mapping, 2, 108
The Living Murray icon site, 68, 106–7, 108
Monica Morgan's descriptions, 57–8, 135–6
Lin Onus painting, *plate [10] between 112–13*, 135, *136*
Barmah–Millewa Forum, 68
basket weaving, 121, 131
Bateson, Gregory, 50, 52
Behrendt, Jason, 132
Bell, Diane, 143
binaries, *see* modern thinking
biodiversity, 29, 34, 39
  wilderness thinking, 22–3
  see also ecosystem destruction
biology, 48
  see also animals; plants
birds, 29, 34, 128
  Barmah–Millewa forest, 57–8, 135; impact of 'rain rejection' flows, 36
  Coorong and Lower Lakes, 1, 2, 121
  Macquarie Marshes, 44, 65
black swan, 121
Blackmore, Don, 30
Blowering Dam, 62–3, *plate [1] between 80–1*
blue-green algae, 36, 60, 120, 127, 136
Bogong moths, 63, 140
Boyle, Robert, 5
Braun, Bruce, 6, 22–3, 108
breeding patterns and events, 1, 34, 36, 44, 121
  fish, 29, 36, 128
  focus of environmental flows, 128
Brennan, Justice G, 75
British Columbia temperate rainforest dispute, 6, 22–3, 25, 67, 108
Bunyip, 122
Burrendong Dam, 44
bush food, 57–8, 131
  Yorta Yorta land use and occupancy mapping work, 2
  see also animals; plants

CAC, *see* Community Advisory Council
Canada, 134
  temperate rainforest conflict, 6, 22–3, 25, 67, 108
cap, 39
carp, 2, 36, 133
Cartesian dualism, *see* modern thinking
Celts, 19
Chowilla Floodplain, 34, 106–7
citizenship, 67, 81, 95–6
civilisation, 20–1, 22–3, 75–6, 96
climate change, 101, 145
co-management arrangements, 73, 87
COAG National Water Initiative (NWI), 40–1, 87–8
cod, *see* Murray cod
colonisation, 3–4, 6–7, 20–2, 31–4, 133
  hierarchical civilisations theory, 20–1, 23, 75–6, 96
  survival, 79–82
commercial fishing, 37
commodification, *see* economic resources
common law, 45
Commonwealth of Australia Constitution, 33
Commonwealth Scientific and Industrial Research Organisation, 101
Community Advisory Council (CAC), 38
  joint meeting with MLDRIN, 68, 120, 126, 127
community values, *see* values
connectivity, 11–15, 27–31, 47–65, 118–48
  land use and occupancy mapping, 2, 23, 107–8
  language of MLDRIN-Commission MoU, 110
  within MLDRIN alliance, 96–8
  proliferation and purification, 16–18, 46
  Psyche Bend Lagoon example, 43–4
  see also continuity principle; country; cultural flows
Connell, Daniel, 40, 41
consent principle, 67–8, 117
Constitution, 33
  MLDRIN, 94–5
continuity principle, 67–8, 117, 122, 135–6
  court judgements on 'tradition', 73–6
  mapping to establish, 23
Coorong and Lower Lakes, 60, *plate [1] between 112–13*, 121
  management planning, 68–9, 85, 106–8
  Matt Rigney's recollections, 1–2, 50–1
  tidal fish traps, 76–8
  wildlife, 1–2, 65; environmental rehabilitation, 106–7
  see also Murray Mouth
Corporations Act 2001 (Cwlth), 55, 93

165

INDEX

Council of Australian Governments National Water Initiative (NWI), 40–1, 87–8
country, 11–15, 33–4, 50–65, 69–70, 80–2, 118–48
    connection with Dreaming, 30, 89, 122
    difference between caring for and natural resources management, 72–3
    The Living Murray icon sites, 106–8
    MLDRIN training courses for young traditional owners, 96
    nations in MLDRIN confederation, 93, 94
    proof of rights required in native title, 74–6
    settlers' perceptions, 3–4, 6–7, 51
    stolen generations, 99
    traditional owners' relationship with, 110–11; speaking for, 94, 102, 112–17, 131
    see also connectivity; native title; natural resource management; places; water management
creation stories, see Dreaming stories
Crew, Jeanette, 72–3, 92–3, 99, 121–2, 131
CSIRO, 101
Cullen, Peter, 115
cultural change, 73–82
    in western thinking, 20–1, 73–6
cultural diversity, 20, 67–79, 82–6, 124
    principles used by state institutions to accommodate, 67–8, 117
    see also continuity principle; recognition principle
cultural (customary) economy, 129–34
cultural flows (Indigenous water allocations), 87–90, 119–29, 145
    Behrendt and Thompson's recommendation, 132–3
    publications about, 81, 87
cultural heritage, 79–90, 100–5
    land use and occupancy mapping, 2, 23, 107–8
    legislation, 54
    MLDRIN funding, 94–5
    see also Indigenous knowledge; native title
cultural identity, see identity
cultural living, 58, 138
    see also bush food
cultural recognition, see recognition principle
cyanobacteria, 36, 60, 120, 127, 136

Daly River, 14
dams, see infrastructure and water storage
Dareton, 92–3
Darling River, 27–8, 36, 127
Dartmouth Dam, 36, 144
Deniliquin, 60, 78
    MLDRIN meeting, 98
    Werai Forest, *plate [4] between 80–1*, 131
Descartes, René, 4
dialogue, see negotiations
dispossession, 57–9, 75, 133
Dreaming, 64, 65, 138
Dreaming stories, 2, 3, 30–1, 55–7, 89, 122
    Ngarrindjeri, 30, 55–6, 86
    Yorta Yorta, 2, 30
drought, 27–9, 34, 36–8, 137
    impact on public debates about water, 130
    policy initiatives in response, 31–2, 40–6, 144–5, 146
    water storage levels, 129; Blowering Dam, 63
'dry inflow', 46
dualism, see modern thinking
ducks, 1, 44, 65
duckweed, 57–8

Echuca, 2, 60
    MLDRIN meeting (March 2004), 72, 109
eco-philosophers, 18
ecological connectivity, see connectivity
ecologically sustainable development, see sustainability
ecologists, 18, 124, 146
    connectivity concept, 47–50
    precautionary principle, 42–3
    resilience concept, 64
economic resources, 3–10, 31–46, 141–2
    cultural economy, 129–34
    cultural flows and, 122–3
    environmental flows and, 126–8, 130
    Tony Peachey's realisation about, 62
    see also agriculture; infrastructure and water storage; sustainability; water trading
ecosystem destruction (ecocide), 18, 26, 34–46, 57–64, 144–8
    acknowledging, 134–41
    see also extinct species
Egler, Frank, 143
Egypt, 5, 17
Emmott, Angus, 45
employment, 62, 107, 131–2
    with water bureaucracies, 72, 92, 110
endangered species, 36
    see also extinct species
Eng, David, 135

England, 19
ENSO, 28
entitlements, *see* water entitlements
Environment Victoria, 101
environmental flows, 39–40, 105–9, 124
　impact of drought, 42, 129
　traditional owners' concerns, 112, 126–8
　see also The Living Murray initiative
environmentalists, 22–3, 24–5, 114–15
ephemeral wetlands and lakes, 34, 45, 136
Eslake, Saul, 40
evaporation, 52, 144
extinct species, 2, 36
　ramifications in terms of identity, 58–9
extraction cap, 39

Farley Consulting Group, 11–13
farming, *see* agriculture
Federal Court, 74–5, 84, 91
federation, 7, 32, 86
'feeling-as-knowing', 143
First Step decision (The Living Murray goal), 105, 107, 144
fish and fishing, 30, 36–7, 82, 133, 136
　Barmah–Millewa forest, 2, 37, 51, 135
　Canada, 134
　Coorong and Lower Lakes, 1, 2, 76–8
　environmental flows and, 127, 128
　extinct and endangered species, 2, 36
　Lower Murray, 51; Agnes Rigney's reminiscences, 58, 122
　Macquarie Marshes, 131
　see also Murray cod
fish traps, 34, 76–8
floodplains, 27, 34, 137, *139*, 143
　Chowilla, 34, 106–7
floods, 27–9, 32, 34, 140
　absent fish species once found during, 2
　Barmah–Millewa forest, 13, 36–7, 127–8; 'rain rejection' flows, 35, 36, 127
　breeding cycles and events, 29, 36, 44
　Lake Eyre Basin, 30
　traditional owners' desire for, 65, 112, 119, 120, 131
flows, 27–37, 38
　'dry inflow', 46
　The Living Murray icon sites, 106–7
　Running the River arrangement, 9–10, 141, 144
　see also cultural flows; environmental flows; water entitlements

forests and forestry, 6, 22–3, 25, 67, 108
　The Living Murray icon sites, 106–7
　river red gum logging, 114
　Werai, *plate [4] between 80–1*, 131
　see also river red gums
Freeman, Ramsay, 62–3, 84, *103*, 123, 128
Frenchman Creek, 10, *11*
Friends of the Earth, 101
frogs, 140, 143

Germany, 6, 19
Gibbs, Leah, 29–30, 45
'government', xii
government agreements, 32–3, 38
　with Indigenous peoples, 85, 101; MLDRIN, 94, 99–100, 101–5, 110
　National Water Initiative (NWI), 40–1, 87–8
　see also The Living Murray initiative
government negotiations, *see* negotiations
government policy, 38–46, 86–90, 144–5
　MLDRIN's involvement, 97–8, 100–1, 107–17
　native title claims, 68–9
　in response to drought, 31–2, 40–6, 144–5, 146
　see also law and legislation; sustainability; water management
Green, Roy, 18–19
groundwater, 33, 39, 43, 52
gums, *see* river red gums
Gunbower and Koondrook–Perricoota forests, 106–7
Gundagai, 32
Gunditjmara people, 34, 115

Hamilton, Tracy, 99
Haraway, Donna, 15, 48, 49, 70
Hattah Lakes, 106–7
heritage, *see* cultural heritage
hierarchical civilisations theory, 20–1, 23, 75–6, 96
High Court, 73–6, 115
Hindmarsh dispute, 84–5
historical narratives, 3
historical time, 21
'history, tide of', 75
holistic knowledge, 11–15, 19
　see also amodern knowledge; connectivity; Indigenous knowledge
Hountondji, Paul, 79
Howard, John, 42

INDEX

Howitt, Richie, 69, 70, 73
Huddleston, Paddy, 14
human beings, 4–5, 17, 55
   in Aboriginal peoples' thinking, 13–15, 51–4, 56, 73
   in Eurocentric notions of wilderness, 22–3
   hyper-separation, 48–50, 73, 76, 130
   in 'natural resources management' definition, 72
   survival, 18, 50, 51–4
   water contained in bodies, 29, 50
human rights language, 76, 81–6
Hume Dam, 35, 144
Hunter, Richard, 50–1, 63–4, 83
hunting, 1, 21, 131
hyper-separation (hyper-incommensurability), 48–50, 73, 76, 130

identity, 56, 74–82, 108, 133
   Hindmarsh dispute, 84–5
   impact of loss of country, 58–9, 61
   political, 70, 95–100, 103–4
   primitivism framework, 22–3
   as traditional owners, 74–6, 79–82, 110–12
   see also national identity
imagery, 30, 37, 137
'Indigenous Basin-Wide Gathering' (Canberra), 97
Indigenous knowledge, 10–15, 20–5, 50–90, 119–48
   'feeling-as-knowing', 143
   hierarchical civilisations theory, 20–1, 23, 75–6, 96
   Latour's classification as 'amodern', 19
   MLDRIN–Commission MoU, 110
   MLDRIN – environmental NGOs, 116
   progressive reason theory, 21, 75, 79
   traditions behind cultural economy, 132
   translation and transformation, 69, 71–6, 104, 109–12, 123–34
   wetland productivity, 33–4, 121
   see also country; Dreaming stories; negotiations; places (sites); premoderns; construction of Indigenous people as 'Indigenous nations', 103–4
Indigenous Partnerships Project, 107–8
Indigenous water allocations, *see* cultural flows
infrastructure and water storage, 7–10, 33, 34–6
   Aswan Dam, 5, 17
   Blowering Dam, 62–3, *plate [1] between 80–1*

Dartmouth Dam, 36, 144
Hume Dam, 35, 144
impact on Macquarie Marshes, 44
impact on Macquarie River, 120
impacts of drought, 40, 63, 129
The Living Murray focus, 108–9
Menindee, 136
Running the River arrangement, 9–10, 141, 144
total water available, 129; percentage allocated for 'environmental use', 126
see also water entitlements
Ingold, Timothy, 50, 52, 55, 56
inlet regulators, 10, *11*
integrated catchment management policy, 39
intellectual traditions, *see* Indigenous knowledge; modern thinking
intercultural engagement, *see* cultural diversity
interest group model in environmental management, 69–70
intergovernmental agreements, 32–3, 38
   National Water Initiative (NWI), 40–1, 87–8
   see also The Living Murray initiative
international rights language, 76, 81–6
interspecies kinship, 121
irrigation, *see* agriculture; infrastructure and water storage

Jackson, Sue, 14, 88
Joachim, Lee, 2
   on Barmah–Millewa lake, 13
   children, *plate [5] between 112–13*
   on protection of Indigenous rights, 81, 82
   relationship with Murray River, 53–4, 61, 119
   understanding of past as part of future, 79–80
   on water flows, 119–20; environmental, 127–8
joint management arrangements, 73, 87
Jones, Henry, 44

Kazanjian, David, 135
Kinnane, Steve, 13, 112
kinship, interspecies, 121
knowledge traditions, *see* Indigenous knowledge; modern thinking
Kumerangk, 84–5

lakes Alexandrina and Albert, *see* Coorong and Lower Lakes
Lake Eyre Basin, 29–30, 45

land tenure system, *see* national parks
land use and occupancy mapping, 2, 23, 107–8
land use changes, impact of, 135–6
Langton, Marcia, 24–5
language, 71–6
    used by Commission, 9–10
    of cultural flows, 89–90, 119–28
    of National Water Initiative, 87–8
    postmoderns' belief, 15–16
    of sustainability, 24–5
    'tangible objects' approach in cultural heritage, 86–7
    of universal rights, 76, 81–6
language groups, 2
language used by MLDRIN, 81, 95
    about cultural economy, 131, 132
    about 'Indigenous water allocations', 125
    in Commission MoU, 99–100, 104, 110
    in environmental NGOs agreement, 113–14
Latji Latji Nation, 93, 94
Latour, Bruno, 4–5, 15, 83, 146
    argument about Indigenous peoples' views of world, 19
    hyper-incommensurability, 48
    purification and proliferation, 16–18, 46
law and legislation, xii, 31–2, 38, 87
    Aboriginal peoples' water issues, 88–9
    constitutional arrangements, 33
    corporations, 55, 93
    definition of 'natural resource management', 71–2
    native title, 22
    nature/culture separation, 54–5, 72;
    common law concepts, 45
    sustainability policy, 24
Lindsay, Biddy, 14
Lingiari Foundation, 87
Littlewood, Roland, 55
The Living Murray initiative, 39–40, 71, 105–9, 128, 130
    community consultation processes, 105–7, 110
    First Step decision (return goal for river health), 105, 107, 144
    icon sites (Significant Ecological Assets), 68, 106–8
    MLDRIN's participation, 94, 107–9
loss, 60–4, 134–41
    cultural living, 58, 138
    western culture's views on, 20–1
    see also ecosystem destruction

Lower Murray, 27–8, 36
    Hindmarsh dispute, 84–5
    Richard Hunter's vision, 63–4
    Pink Lake, *plate [2] between 112–13*, 141, *142*
    Agnes Rigney's descriptions, 50–1, 60; Mulyewongk, 122
    see also Coorong and Lower Lakes

Mabo decision, 75–6, 83, 110
Macleans Beach, 78
Macquarie Marshes, 44, 65, 131
Macquarie River, 120
Main, George, 147
mapping, 2, 23, 107–8
market economy, *see* economic resources
market mechanisms, *see* water trading
mining of river sand, 2
Ministerial Council, *see* Murray–Darling Basin Ministerial Council
Mitchell, Junette, 133, 136
Mitchell, Timothy, 5, 15–16, 17, 21, 146
Mitta Mitta River, 36
MLDRIN, *see* Murray Lower Darling Rivers Indigenous Nations
'modern', xii
modern thinking, 3–10, 15–25, 67–79, 146–8
    complexity and, 14
    constructs used in natural resources management, 72, 116
    ecology/economy dualism, 24–5, 129–34
    environmentalists, 22–3, 24–5, 114–15
    hyper-separation, 48–50, 73, 76, 130
    Indigenous identity constructs and, *see* identity
    international rights discourse and, 76, 81–6
    in legislation, 54–5, 72
    perceptions of cultural change/loss, 20–1, 73–6
    postmodern critiques, 15–16, 21
    in water management, 7–10, 31–46
    see also amodern knowledge; economic resources; premoderns; science
Moira Channel, *32*, 36
Morgan, Monica, 2, 57–8, 92, 96–7
    on acknowledging loss, 135–6, 140, 147, 148
    employment with Commission, 72, 110
    realisation of difference between natural resource management and caring for country, 72
    on UN approach to human right to water, 53

INDEX

Morrison, Joe, 88
mouth of Murray, *see* Murray Mouth
Muecke, Stephen, 133
Mulyewongk, 122
Murray cod *(pondee)*, 2, 30, 36
   commercial catch, 37
   Ngarrindjeri creation story, 30, 55–6
Murray crayfish, 36, 37
Murray–Darling Basin Agreement, 38
Murray–Darling Basin Authority, 38
Murray–Darling Basin Commission, 7–10, 38, 71, 91
   Barmah–Millewa Forum, 68
   Blackmore, Don, 30
   Farley Consulting Group engagement, 11–13
   Green, Roy, 18–19
   'Indigenous Basin-Wide Gathering', 97
   The Living Murray consultations, 105–6
   Running the River arrangement, 9–10, 141, 144
   water buyback scheme, 42, 46
Murray–Darling Basin Commission and MLDRIN, 92, 97
   funding arrangements, 94, 96
   interpretation of 'natural resource management', 72, 73
   The Living Murray initiative, 94, 107–9, 110
   memorandum of understanding, 99–100, 101, *102*, 103–5, 110
   'Murray–Darling Basin Environmental Resources Study', 86–7
Murray–Darling Basin Initiative, 38–9, 89
Murray–Darling Basin Ministerial Council, 38–40
   see also Community Advisory Council; The Living Murray initiative
Murray Lower Darling Rivers Indigenous Nations (MLDRIN), xi–xii, 91–117
   agreements with other organisations, 99–100, 101–5, 110; non-governmental, 101–2, 113–14, 116
   Executive Officer, *see* Ross, Steven
   funding, 94–5, 96
   interviews with traditional owners involved with, xii, xvi
   publications, 81, 97, 100, 105–6
   strategic plan, 109, 125
   see also Murray–Darling Basin Commission and MLDRIN

Murray Lower Darling Rivers Indigenous Nations (MLDRIN) meetings, xii, 11–12, 94, 97–8
   Dareton (July 2001), 92–3
   Echuca (March 2004), 72, 109
   Swan Hill (September 2003), 53
   Tumut (December 2003), 62–3
   Wagga Wagga (with CAC, July 2005), 68, 120, 126, 127
Murray Mouth, 34, 38, 40, 106–7
   Farley Consulting Group report on Aboriginal concerns, 12
Murray Waters Agreement, 32–3
Murrumbidgee Catchment Management Authority, 89, 101
Murrumbidgee Irrigation Area, 33
Murrumbidgee River, 28, 32, plate [3] between 112–13
Murrumbidgee Water Sharing Plan, 89
mussels, 36, 95
Mutti Mutti Nation, 93, 94, 101
   Crew, Jeanette, 72–3, 92–3, 99, 121–2, 131
   see also Pappin, Mary
mutual recognition, *see* recognition principle

Nari Nari Elders, 101
narratives, 1–25
   biological discourse, 48
   western rights discourse, 76, 81–6
   see also language; modern thinking; primitiveness
national identity, 104
   Murray–Darling Basin in, 3, 30, 37
National Indigenous Advisory Committee, 87
national parks, 114
   joint management arrangements, 73, 87
   'wilderness thinking', 22–3
National Parks Association of NSW, 101, 114
National Water Initiative (NWI), 40–1, 87–8
'nations', 103–4
native fish, *see* fish and fishing
native title, 73–6, 86, 91, 110–11
   community dialogue, 83
   government policy about, 68–9
   influence of traditional owners in government policy resulting from recognition, 87–9, 101
   Yorta Yorta decision, 73–5, 76, 91
natural resource management, 71–3, 119, 144–5

MLDRIN involvement, 94, 100–1, 110, 112–17
  see also cultural flows; environmental flows; sustainability
natural resource use, *see* economic resources
nature and society binary, *see* modern thinking
Nature Conservation Council of New South Wales, 101–2
negotiations, 66–90, 95–117, 138
  about cultural economy, 131
  about cultural flows, 89–90, 124–9
  establishing MLDRIN, 92
  see also language
New South Wales, 32–4, 37, 38, *plates between 80–1, plates between 112–13*
  Blowering Dam, 62–3, *plate [1] between 80–1*
  Darling River, 27–8, 36, 127
  Frenchman Creek inlet regulator, 10, *11*
  Macquarie Marshes, 44, 65, 131
  recognition of Aboriginal peoples' water issues, 88–9, 101
  river red gums, *plates [3–4] between 80–81, plate [7] between 112–13*, 114, 138
  Yorta Yorta petition, 74
  see also Barmah–Millewa; Mutti Mutti Nation; Wiradjuri Nation
New South Wales Department of Land and Water Conservation, 92, 101
New South Wales National Parks Association, 101, 114
New South Wales Parks and Wildlife Service, 131
New South Wales *Water Management Act 2004*, 88–9
Ngarkat people, 141
Ngarrindjeri Nation, *93*, 94, 101
  battle at Pink Lake, 141
  creation story, 30, 55–6, 86
  'feeling-as-knowing', 143
  Hindmarsh dispute, 84–5
  Hunter, Richard, 50–1, 63–4, 83
  native title dialogue, 83
  ngatjis, 121
  Rigney, Agnes, 50–2, 58, 60, 97, 122
  see also Coorong and Lower Lakes; Lower Murray; Rigney, Matt
ngatjis, 121
Ngunawal Elders, 101
noble savage concept, 25
  see also primitiveness

North America, 58–9, 104
  see also Canada
North Central Catchment Management Authority, 101
North West Nations, 101
Northern Territory, xii, 14, 119
Nuu-chah-nuth people, 22–3, 25
Nyampaa Nation, 97–8

occupancy and land use mapping, 2, 23, 107–8
Oldfield, Sharon, 30, 45
Olney, Justice HW, 74–5
Onus, Lin, 'Barmah Forest' painting, *plate [10] between 112–13*, 135, *136*
Onus v Alcoa, 115, 124
ownership of water, 33

Pappin, Mary, 57, 65, 82, 95, *99*
  on carp, 133
  on cultural heritage, 59, 80–1
  on sand mining, 2
  on water flows, 120; environmental, 112–13, 126, 128
Paz, Octavio, 85
Peachey, Tony, 44, 62, 65, *103*, 120, 131
pelicans, 1, 2, 135
perch, 29, 36
philosophies, *see* Indigenous knowledge; modern thinking
Pink Lake, *plate [2] between 112–13*, 141, *142*
place, 29–30, 146–7
  Haraway's 'situated knowledge' argument, 49
  holistic approaches in western knowledge traditions, 19
  mapping, 6, 23, 107–8
  in modern thinking, 21
  sentient ecology, 50
  see also country
'place-bereavement', 61
places (sites), 86–7, 122
  in and around Blowering Dam, 62–3
  High Court *Onus* judgement, 115
  Hindmarsh dispute, 84–5
  The Living Murray icon sites, 106–8
  The Living Murray icons (Significant Ecological Assets), 68
  Macquarie Marshes, 131
  middens, *plate [9] between 80–1*, 95
  Ngarrindjeri–Ngarkat battle, 141

INDEX

Mary Pappin's thoughts on protecting, 59, 80–1
plants, 13, 62
  adaptation to variability, 29
  for basket weaving, 121, 131
  biological discourse, 48
  colonialists' perceptions, 6
  cultural flows and, 89, 120, 121
  duckweed, 57–8
  hollow tree habitats, 72
  land use and occupancy mapping, 2, 23, 107–8
  ramifications of extinction in terms of identity, 58–9
  see also agriculture; forests and forestry
Plumwood, Val, 18, 48
policy, see government policy
political identity, 70, 95–100, 103–4
pondee, see Murray cod
population, 26
  native fish, 36
postmoderns, 15–16, 21
Povinelli, Elizabeth, 78
precautionary principle, 42–3
premoderns, 4–5
premoderns, construction of Indigenous people as, 19
  in environmentalist narratives, 23, 25
  in negotiations, 90
  perception of cultural change, 20–1
primary industries, see agriculture; forests and forestry
primitiveness, 6, 22–3
  hierarchy of civilisations theory, 20–1, 23, 75–6, 96
  noble savage concept, 25
  progressive reason theory, 21, 75, 79
profileration and purification, 16–18, 46
progressive reason model, 21, 75, 79
Psyche Bend Lagoon, 43–4, *plate [4] between 112–13*
purification and proliferation, 16–18, 46
Pyle, Robert, 59, 85

quality of water, see water quality
Queensland, 27, 28, 38, 39

rainfall, 27–30, 140
  dependence of cultural flows on, 129
  'rain rejection' flows, 35, 36, 127
  see also drought; floods

Read, Peter, 61
recognition principle, 67–8, 70, 85–90, 102, 117
  Indigenous response to The Living Murray, 106
  language used in claims, 71–6
  to speak for country, 113, 115
  see also native title
reconciliation, 85, 133
recreational fish catches, 37
red gums, see river red gums
'regenerative agriculture', 147
representations, 17, 67, 141–5
  political–legal arrangement, 86
  postmoderns, 15–16
  universal knowledge, 48–9
reservoirs, see infrastructure and water storage
resilience, 64–5
resource management, see natural resource management; water management
resource use, 13–14, 24–5, 33–4
  language of MLDRIN–Commission MoU, 104
  see also bush food; economic resources
respect for country, see country
Rigby, Kate, 19
rights language, 76, 81–6
Rigney, Agnes, 50–2, 58, 60, 97, 122
Rigney, Matt, 55–6, 65, *102*, 108, 141
  on cultural flows, 121
  on degradation of country, 60
  on environmental flows, 126
  reminiscences, 1–2, 50–1
  on tidal fish traps, 76–7
Rigney, Peter, 1
River Murray Commission, 32
River Murray Water, 7–10
  see also infrastructure and water storage
river red gums, 29, 30, *plates [3–4] between 80–81*, *plates between 112–13*, 137, 138, 139
  MLDRIN campaign, 114–15
  see also Barmah–Millewa
river sand mining, 2
Robin, Libby, 6–7, 72, 146–7
romanticism, 19
Rose, Debbie, 14, 49, 54, 143
  on dialogue, 70, 116
  on moderns, 69
  on scholars, 84
Ross, Steven (MLDRIN executive officer), 99, 109, 116

childhood experiences, 78
on cultural flows, 123, 126, 128
description of Dareton meeting, 93
paper on MLDRIN, 97
river red gum campaign, 114–15
Rousseau, Jean-Jaques, 25
run-off, 27, 129
irrigation water, 43–4
Running the River arrangement, 9–10, 141, 144

Sahlins, Marshall, 79
salinity, 27, 38, 52
Coorong waters, 121
Pysche Bend Lagoon, 43–4
sand mining, 2
scale-making, 96
science, 4–10, 19, 48–50, 115, 141–4
advice on returns for river health, 40, 106
anthropology, 18, 20–1, 55
fish species, unrecorded, 2
MLDRIN partnerships, 101–2, 116
precautionary principle, 42–3
proliferation and purification, 16–18, 46
see also technology
sediment loads, 35–6
sentience, 49–65
settler societies, 20–2, 67–86, 95–6, 133
settlers and settlements, 3–4, 6–7, 31–4, 51, 137
Sillitoe, Paul, 5–6, 55
Sinclair, Ian, *102*
Sinclair, Paul, 137
sites, *see* places
'situated engagement', 70–1
'situated knowledges', 49
Smith, Ben, 49, 73, 82, 113
society and nature binary, *see* modern thinking
soils, 6, 7, 27
South Australia, *28*, 32–3, 37, 38, 144, 148
Chowilla Floodplain, 34, 106–7
native title policy, 68–9
see also Coorong and Lower Lakes; Lower Murray; Murray Mouth; Ngarrindjeri Nation
Southern Oscillation, 28
space, 20–2
see also place
speaking for country, 94, 102, 112–17, 131
spiritual narratives, *see* Dreaming stories
stakeholder model in environmental management, 69–70
stolen generations, 99, 112
Strathern, Marilyn, 54

subjectivity, 143
Indigenous peoples' framework, 76–9
sentience, 49–65
see also identity
Suchet-Pearson, Sandie, 69, 70
survival, 18, 50
of Indigenous culture, 79–82, 90
Indigenous knowledge about, 51–4, 122
see also ecosystem destruction
sustainability, 19, 38–46, 146–7
ecology/economy dualism, 24–5, 129–34
forestry principle, 6
see also The Living Murray initiative
Swan Reach mission, 51, 64, 122
swans, 57–8, 121

Taungurung Nation, 93, 94
technology, 10, 42
Indigenous people in outback using, 21
see also infrastructure and water storage; science
terra nullius, 75–6
thinking systems, *see* Indigenous knowledge; modern thinking
Thompson, Peter, 132
threatened species, 36
see also extinct species
time, 20–1
Dreaming, 64, 65, 138
see also continuity principle; premoderns
Tobias, Terry, 108
totemic systems, 121
tourism, 131, 137
trade, 131
see also water trading
'tradition', 73–6
see also cultural change; Indigenous knowledge
traditional identity, *see* identity
traditional knowledge, *see* Indigenous knowledge
'traditional owners', xii
translation and transformation, 69, 71–6, 104, 109–12, 123–34
trees, *see* forests and forestry; plants
Tsing, Anna, 21, 82–3, 96, 105
Tully, James, 67–8
Tumut, 62–3

umwelt, 49
United Nations, 53, 81

universal narratives, *see* modern thinking
universal rights language, 76, 81–6
use of resources, *see* resource use

values, 6, 19, 45
  constraining environmental flows, 126–7
  Indigenous peoples', 12–15, 24, 68, 101, 104;
  recognition by government, 85, 86–9
variability, 27–9, 31–2, 34, 146
  cultural flows and, 123
  environmental flows and, 127
vegetation, *see* plants
Victoria, 28, 31–3, 36, 38, *plates [5] and [9]
  between 80–1*
  Gunditjmara people, 34, 115
  North Central Catchment Management Authority, 101
  Psyche Bend Lagoon, 43–4, *plate [4] between 112–13*
  river red gums, 114
  see also Barmah–Millewa; Yorta Yorta Nation
Victoria River, *plate [8] between 112–13,* 137, *139*
Victorian Environment Assessment Council, 114
Victorian National Parks Association, 101
von Doussa, Justice J, 84

Wadi Wadi Nation, 93, 94
Wagga Wagga, 33–4
  MLDRIN joint meeting with CAC, 68, 120, 126, 127
Wamba Wamba Nation, 93, 94, 99
  woven grass baskets, 131
  see also Ross, Steven
water entitlements (allocations), 7, 32–3, 37, 39–42
  extraction cap, 39
  governments' underlying assumption, 43
  Running the River arrangement, 9–10, 141, 144
  voluntary buyback scheme, 42, 46
  see also cultural flows; environmental flows; water trading
water flows, *see* flows
water management, 7–10, 18–19, 26–46, 118–48
  impact on cultural living, 58

  Indigenous peoples' participation, 68–9, 72, 85, 86–90, 97–8
  see also infrastructure and water storage; Murray Lower Darling Rivers Indigenous Nations
Water Management Act 2004 (NSW), 88–9
water quality, 35–7, 38–40, 50–1, 60–1, 65
  Barmah–Millewa forest, 36–7, 60, 127, 135
  Canada, 134
  Macquarie River, 120
  Menindee, 136
  see also salinity
water trading, 45
  cultural flows, 132–3
  'dry inflow', 46
  The Living Murray initiative, 108–9
  National Water Initiative, 41
  water prices, 41, 126–7, 130–1
waterbirds, *see* birds
watertable, 33, 52
weather, *see* rainfall
weaving, 121, 131
weirs, *see* infrastructure and water storage
wells, 19
Werai forest, *plate [4] between 80–1,* 131
Wergaia Nation, 93, 94
western knowledge traditions, 3–10, 15–25
  see also amodern knowledge; modern thinking
western rights discourse, 76, 81–6
wetlands, 33–4, 140–1
  cultural flows and, 121, 129
  environmental flows and, 126
  ephemeral, 34, 45, 136
  The Living Murray icon sites, 106–8
  Macquarie Marshes, 44, 65, 131
  Murrumbidgee Water Sharing Plan, 89
  Psyche Bend Lagoon, 43–4
  Running the River arrangement, 144
  see also Barmah–Millewa; Coorong and Lower Lakes
wilderness, 22–3
Wilderness Society, 101
wildlife, *see* animals; plants
Williams, Robert, 81
Wiradjuri Nation, 32, 93, 94, 101
  Freeman, Ramsay, 62–3, 84, *103,* 123, 128
  Peachey, Tony, 44, 62, 65, *103,* 120, 131
<www.mldrin.org.a>, 95

Yellowin Bay, 63, *plate [1] between 80–1*
Yorta Yorta Nation, 61, 91–2, 93, 94
    Barmah–Millewa Forum, 68
    Dreaming stories, 2, 30
    land use and occupancy mapping, 2, 108
    management agreements, 85, 101
    native title decision, 73–5, 76, 91
    see also Atkinson, Henry; Barmah–Millewa; Joachim, Lee; Morgan, Monica